ROBOTICS

Sensors and the Environment

Ian Chow-Miller

Cavendish Square
New York

Published in 2017 by Cavendish Square Publishing, LLC
243 5th Avenue, Suite 136, New York, NY 10016

First Edition

Website: cavendishsq.com

Names: Chow-Miller, Ian.
Title: Sensors and the environment / Ian Chow-Miller.
Description: New York : Cavendish Square, 2017. | Series: Robotics | Includes index.
Identifiers: ISBN 9781502619389 (library bound) | ISBN 9781502619396 (ebook)
Subjects: LCSH: Detectors--Juvenile literature. | Transducers--Juvenile literature.
Classification: LCC TK7872.D48 C46 2017 | DDC 681'.2--dc23

Editorial Director: David McNamara
Editor: Fletcher Doyle
Copy Editor: Nathan Heidelberger
Art Director: Jeffrey Talbot
Designer: Alan Sliwinski
Production Assistant: Karol Szymczuk
Photo Research: J8 Media

The photographs in this book are used by permission and through the courtesy of: Duncan Chard/Bloomberg via Getty Images, cover, 1; KAREN BLEIER/AFP/Getty Images, 4; Ian Chow-Miller, 17, 23, 27, 32, 38, 39, 41, 42, 44, 47 (3), 51 (2), 61, 62-63, 65, 66, 69, 71, 75, 78, 80-81, 83, 85, 86, 90, 92, 93, 94, 97, 105, 107, 112, 114; Photo used with permission. ©2016 The LEGO Group, 9; Vex Robotics, 11; Credit: Bruno Ismael Silva Alves/Shutterstock.com, 14; ZUMA Press Inc/Alamy Stock Photo, 19; Esignua/Shutterstock.com, 20 (top), Ociacia/Shutterstock.com, 20 (bottom); Bas Nastassia/Shutterstock.com, 26; Sheng Li/Reuters/Newscom, 29; Higyou/iStockphoto.com, 34; Julia Tim/Shutterstock.com, 36; ra3rn/Shutterstock.com, 52; Jonathan Daniel/Getty Images, 56; AP Images, 82; Goran Bogicevic/Shutterstock.com, 88; Demeshko Alexandr/Shutterstock.com, 91.

Printed in the United States of America

Contents

A self-driving car uses hundreds of sensors to navigate safely around the roads and highways of our country.

1 Making Sense of Sensors

Stand up and walk toward a wall, then come back to this book. Did you stop before you hit the wall? I bet most of you did. Why? Probably because you saw the wall and you didn't want a nasty bump on your forehead. This is common. Humans use senses like sight all the time to give us information that we then process in order to make decisions. A previous experience of walking into a wall or other solid object told you that it was a bad idea—it probably hurt a lot the last time. So you stopped. Sensing that wall with your eyes and stopping was a smart decision and one that you made immediately based on experience.

Robots use **sensors**, too, from the small hobby or education robots you may be using in the classroom or at home to the million-dollar robots created by tech companies or the military. How does a search and rescue robot crawl through the rubble and find people buried under buildings in an earthquake? Sensors. **Infrared sensors** can detect a person's body heat, and sound sensors can detect noise. How does a **welding robot**

know if it's overheating a piece of metal in an automobile factory? It uses a **temperature sensor**. And how does a car drive several hundred thousands of miles over the last five years with nobody behind the wheel (actually, with no steering wheel at all)? Sensors. The tech company Google has designed and tested software that allows cars to be self-driving. You may even have stood flabbergasted on the side of the road as one of these drove past.

Google's purpose for designing a self-driving car may seem idealistic, but it is also very practical. As the company posted on its website: "Imagine if everyone could get around easily and safely, regardless of their ability to drive. Aging or visually impaired loved ones wouldn't have to give up their independence. Time spent commuting could be time spent doing what you want to do. Deaths from traffic accidents—over 1.2 million worldwide every year—could be reduced dramatically, especially since 94 percent of accidents in the US involve human error."

Achievement of these lofty goals and development of the cars that can make them come true requires the use of sensors. Most notably, Google's self-driving cars use a **laser** mounted on top that shoots and reads the return pattern of sixty-four beams to create a detailed 3-D picture of its surroundings. It compares this data with detailed maps—like those you have on your phone's **GPS**—to help navigate around the highways and local streets alike. Other sensors Google mounts on its self-driving cars are four **radars**, a camera, a GPS, an **inertial measurement unit**, and a wheel **encoder**.

Not many high-school students will be firing lasers tomorrow, but with a little bit of help and by following the information in this book, you will be able to program and use some of the most common sensors out there. Possibly even the same ones Google **engineers** began **programming** with.

Your Robot Is Not Human

Before we explore sensors and how to use them, it's important to understand two basic differences between robots and humans: experience and self-preservation. Your decision not to walk into the wall at the beginning of this chapter came from a sense of self-preservation—the desire not to have harm come to your body. Robots do not have self-preservation. Nor do they have experience the way you and I think of the term. If you program your robot to drive forward and stop when it senses a wall one time, the next time it senses a wall it will crash right into it unless you tell it to stop again. It sounds funny but I always tell my classes, "Your robots are stupid—fast, but stupid." This is not really true or fair to the robots, but it accurately describes how robot behavior can appear to a human. A robot can acquire thousands of pieces of information about its environment through its sensors in a second—more accurate and detailed information than you or I ever could—and it will still drive off a cliff or into a wall unless you tell it not to.

A third and final difference to keep in mind, and one related to experience, is that robots are not really stopping at a wall. They are stopping (if you remember to tell them to) when they sense something using a very particular sensor. For example, a distance sensor (sometimes called an **ultrasonic sensor**) may tell them an object is 4 inches (10 centimeters) away, but the robot doesn't know if that object is a wall or a vase of flowers. And again, it won't stop (wall or vase) unless you tell it to.

Keeping these three differences in mind—robots have no sense of self-preservation; they don't learn from experience the way you and I do; and their lack of experience does not allow them to interpret information, only to read the exact information their sensors give them—it's time to take a look at the most common sensors available for the most common robotics systems, how they work, what we can do with them, and how to use them.

Analog vs. Digital

All the sensors you will ever use, those described in this book and others, will fall into one of two categories: analog or digital. We'll discuss digital first, as those are the easiest to understand and use. A **digital sensor** has only two positions or states. These are commonly referred to as "on" and "off." What this means is a **voltage** is either sent from the sensor to the robot or it isn't. The amount of voltage does not matter; some voltage or zero voltage are the only options. If voltage is received, the robot reads an on signal, and if voltage is not received, the robot reads an off signal. Again, what the robot does when receiving these signals is up to you the programmer.

The most prevalent type of digital sensor is one that senses a touch of some kind. When the sensor is touched, or in some cases pushed, an electronic **gate** is opened and a voltage signal is sent to the robot. Touch sensors are simple to understand and use, and because of this they are very prevalent in the world of educational robotics. While the design may differ slightly in most common sensors, they are all basically the same. In their natural position, as seen below, they are open; no voltage, and therefore no signal, is being sent to the robot. The robot reads zero voltage and translates that as "no touch," or off. When they are pushed, a signal, a certain amount of voltage, is sent to the robot and the robot reads that signal and translates it as "touch."

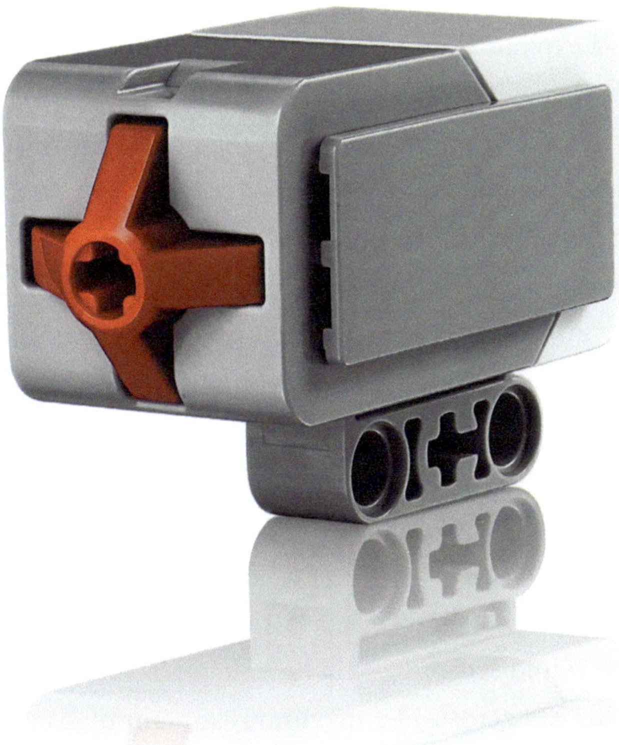

LEGO Education's touch sensor is a digital sensor that is naturally in the off position. It registers a touch when the red button is pushed..

You will notice that each version of the sensor has the ability to be pushed in (an internal spring will pop it back out). While these sensors have a similar design to them, other companies have gone in different directions with theirs. The VEX bump sensor and **limit switch** are digital sensors that detect a push, but they operate differently. The bump switch is more rugged than the limit switch and is generally used when your robot needs to know if it has hit an obstacle. It also makes a great on/off switch for starting a program. The limit switch can be put to good use for "softer" applications, such as detecting when a gate or a door has reached an open or closed position.

We'll explore all these different types of touch sensors in depth in the later chapters.

Let's turn our attention now to the other type of sensor, the analog. An **analog sensor** is one that will return a range of values. (Some people would argue that a sensor is only analog if it returns an infinite range of values, but that is splitting hairs over semantics that are beyond the scope of this book.) An analog sensor will return a specific amount of voltage in a given range to indicate specific information about its environment. To illustrate this information, we'll study one of my favorite sensors, the LEGO Mindstorms NXT sound sensor, or sound sensor for short.

Simply put, this sensor senses sound. Absolute silence will cause the sound sensor to send no voltage to the robot and the robot will read zero, and a maximum amount of sound will cause the sensor to send a predetermined amount of voltage to the robot and the robot will read one hundred. (In actuality, the sensor is reading raw values of **decibels** that are then converted to a scale of 0–100 for easy use and programming in LEGO's NXT or EV3 drag-and-drop programming environments. Were you to use the sensor with a text-based language like **RobotC**, you would have to scale the values yourself or just use the raw decibel values in your program.) Anything between absolute silence and the maximum decibel level this sensor can detect will return a set amount of voltage that the robot will then read and convert into a number between one and one hundred.

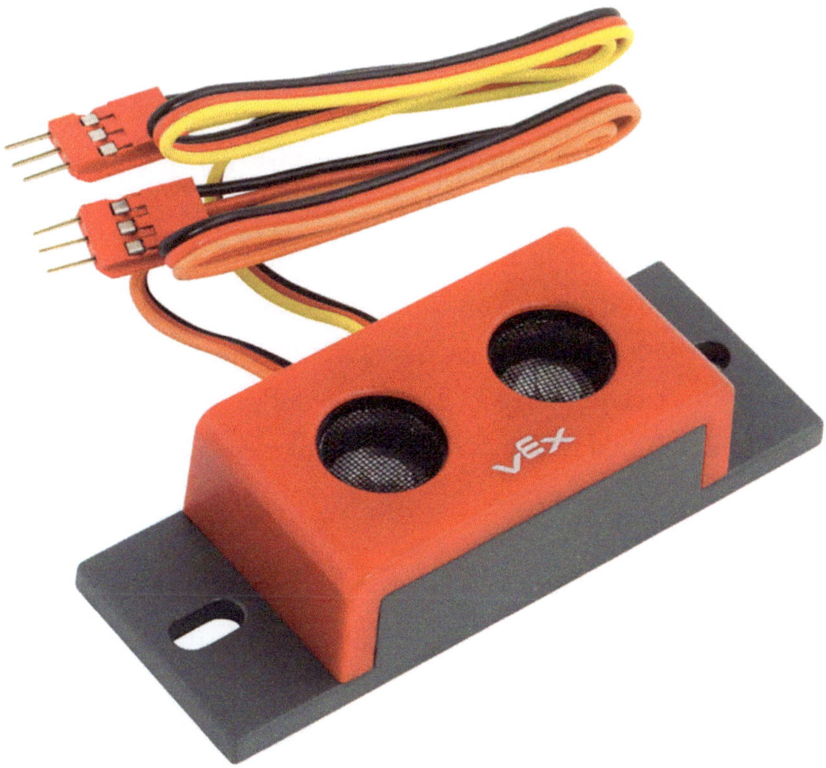

The VEX EDR Ultrasonic Range Finder can measure distances to a maximum of 115 inches (292 centimeters).

Because the robot does not distinguish between you yelling at the top of your voice and a jet plane engine roaring, it is not a true analog sensor, but for most beginning robotics purposes, we can call the sound sensor with its range of values analog, whereas we call the touch sensors with their on/off positions digital.

The path from programming a small robot to respond to sound and a self-driving car traveling the streets of America is not as great as you might think. Engineers do not start by programming an **array** of sixty-four roof-mounted lasers. They start at the beginning, with a basic robot and a basic sensor. And you'll find that some teachers are very fond of their sound sensors. They have a number of really great applications. When running a class project, I can have all the robots lined up and my students absolutely silent so we can start their robots with a loud yell of the word, "Go!" I also have students make their robots go when they hear clapping and stop when it's silent. It is amusing to watch twelve-year-old boys and girls clapping at their robots while they follow them around the classroom. As I watch them clapping away, I always wonder which ones will build and design a new cancer-detecting sensor in the field of medical robotics, or who might build a robot that can detect and defuse a bomb by relaying sensor information to a remote operator. I muse on who might take the next step and become the lead engineer on the self-flying plane of the future. It might even be you. And hopefully this book will help you get started.

In the next chapter of this book, we will look at all of the commercially available sensors for the most popular robotics platforms out there: LEGO Mindstorms and VEX. We will also examine **third-party sensors**—those produced by other companies for those same platforms. Chapter three will explore how to properly build with these sensors and how to write basic programs to use them in the simplest fashion. Chapter four will introduce you to advanced programing, using "if, then" statements (sometimes called forks or **switches**) and "while" statements (sometimes called loops). Finally, chapter five will take all of this information into advanced areas of use. If you get lost with any of the terms in this book, there is a glossary at the back to help you recall what they mean. So let's go and start using our sensors!

Wall-E's eyes look just like common ultrasonic sensors, although his eyes were able to move. Your sensors will be stationary.

2 Selecting a Sensor

A self-driving car can be outfitted with the most expensive sensors operated by the most sophisticated computer system ever, and it will still crash into a wall if it is using the *wrong* sensors, has them set up in the wrong way, or is reading the information from them incorrectly. This chapter is going to help you understand what specific information you can get from sensors, and perhaps even more importantly, what information you can't get. It will discuss which sensors are the correct ones to use in given situations, and how to understand the information you are receiving from the sensors.

Touch (Bump) Sensor

The touch sensor is the most common sensor in educational robotics kits. The reason for this is pretty simple: the touch sensor is a digital sensor. It only has two states: on and off. This makes it easy to use, program, and understand. The most basic thing a touch sensor can tell you is if it has been pushed in or not. If it has been pushed in, it will output a signal saying so; if it hasn't, no signal will be output.

Students competing in robotics competitions often think using a touch sensor will be a great way to stop their robots from crashing into walls and other obstacles. They usually find the opposite is true. Close your eyes and picture this for a moment. Your car is driving along at 30 miles per hour (48 kilometers per hour) and is approaching a wall. A split second before you smash into the wall a bumper-mounted touch sensor hits it and lets the car's operating systems know. If you are lucky, your car engages the brakes immediately, the airbags deploy, and at worst, you have whiplash. If the brakes do not engage quickly enough, the car smashes into the wall and the injury to your body can be a lot worse. It is for this reason that front-mounted touch sensors are not ideal for stopping vehicles from crashing into walls when driving fast.

However one of the most common uses of a touch sensor is for bumping into things—at speeds much slower than 30 miles per hour, in robots a bit smaller than self-driving cars. For example, in the 2015 FIRST LEGO League challenge, Trash Trek, teams may have wanted to know when their robot was lined up directly in front of a recycling machine. It's not feasible to use an ultrasonic sensor for this, but a light bump with the touch sensor can do the job. You can see how a team set up a touch sensor for this purpose on the next page.

Touch sensors have a small point of contact. The robot in the image on the right has an **axle** connected to the touch sensor with a tire on the front to extend the reach of the sensor. Let's go back to our self-driving car example for a moment.

A touch sensor with a small contact point attached to a large tire via an axle can tell you when your robot is aligned properly.

If you have a touch sensor placed at one point on the bumper and you hit an object with a different part of the bumper, your sensor is useless. It's better to extend its range in most cases.

In summary, the touch sensor is better at helping you locate something with a light touch, as opposed to stopping your robot from crashing into obstacles. But there are many other uses for this sensor, such as anytime you have a robot on which you want to push a button to activate something. For example, one of the simplest uses of a touch sensor is as a start mechanism for a program. By simply attaching the sensor somewhere prominent on the robot, you can push it any time you wish to start or stop a program. If you attach a long wire to the touch sensor, you can use it to take a robot vehicle for a "walk," pushing the touch sensor as you start and stop the robot. The touch sensors also make good controls for turning your robot into a video game controller for controlling another robot via Bluetooth. The possibilities are endless, from recording drops of M&Ms in a candy sorter to registering the arrival of a new piece on a simulated factory assembly line.

TOUCH SENSOR TIP

You can use a touch sensor when it is let out, not pushed in. For example, you can place one on the bottom of a wheeled robot vehicle so that the floor is pushing the sensor in. As the robot drives toward where there is no floor, such as the edge of a table, the sensor pops out and the robot knows it is at the edge of the table. In the same way, you can follow the edge of a wall with the sensor pushed in until the wall turns.

We also need to remember that while most touch sensors are of the "bump" kind, VEX makes a limit switch for its EDR line. This sensor, with its thin aluminum arm, is a little more delicate than most touch sensors. It is most definitely not suited for situations where you would bump or crash a robot into something. It is better utilized when you need to recognize a soft, exact touch. My students have put a limit switch to good use when stopping a tollbooth gate once it reaches the down position or a drawbridge when it reaches either its open or its closed position.

Pop-culture robots such as Johnny Five have eyes that see; ultrasonic sensors see objects and detect how far away they are.

Ultrasonic Sensor

The ultrasonic sensor makes it easy for us to understand what it does in that in most cases it is shaped to look like a pair of eyes, similar to many pop-culture robots like Johnny Five and Wall-E.

The ultrasonic sensor has many applications for the robotics student. The simplest thing to do with an ultrasonic sensor is to have the robot drive and stop a certain distance from a wall or other solid object that it "sees." If you recall at the beginning of this book, I asked you to walk toward a wall. The majority of you probably stopped. That's how we began our exploration of sensors, and from there we discussed real life examples of robots that use sensors—like our self-driving car. We already have learned that it is not very convenient to try to stop crashing into an object by placing a touch sensor on the

EYES THAT HEAR

The design of an ultrasonic sensor evokes the eyes of these robots, but it isn't totally accurate to say that the ultrasonic sensor "sees." These sensors are more accurately referred to as distance sensors because what they do is sense the distance an object is away from the robot.

To do this, the sensors send out, or shoot, a high-frequency sound wave, better known as an ultrasonic wave. This wave travels at a known speed, and when it encounters a solid surface, it bounces back at the same speed. Based on how quickly the wave comes back, the sensor can tell how far away the object is. This is the same as **echolocation** used by animals like bats.

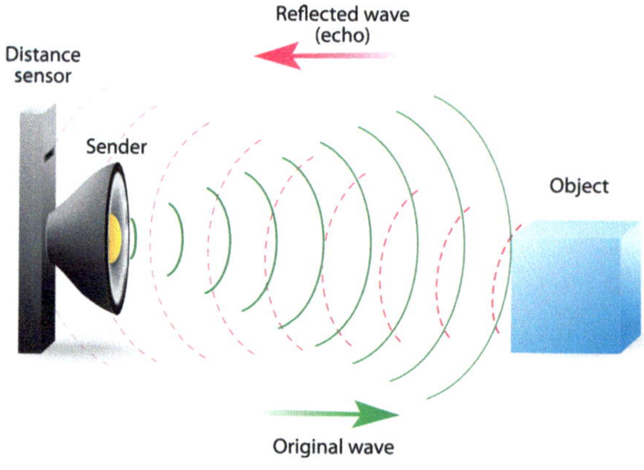

Ultrasonic sensors use one eye to send sound signals and another to receive returning signals.

front of a vehicle. At high speeds, there is not enough time for a car to stop between the time it takes for a bumper-mounted touch sensor to sense a collision and for the sensor to relay that information to the car. But with an ultrasonic sensor, you have a much larger margin for error; it can detect things early enough so that robots have time to stop before a collision. And if you were to mount multiple ultrasonic sensors facing forward, you could protect your car from a collision coming from the front or sides. And it doesn't stop there.

While you definitely don't want to get too close to any objects in front of you, you also don't want to get too far behind a car as well. You always want to maintain a safe distance. Once you know that distance, you can program your car to slow down when it is too close to an object, but you can also have it speed up when it is too far away from a car it was following. The ultrasonic sensor can be used to create a robot that does the same thing. A student recently created a penguin that happily made penguin noises when its mom (a stuffed animal) was near, but started to follow anytime the mom moved too far away. It's not that hard to make the leap from understanding how to build and program a robotic animal with LEGOs to working on self-driving cars.

The ultrasonic sensor doesn't work only along a horizontal plane. Let's pretend you have a rescue robot that needs to cover unsteady terrain in order to reach an injured victim. One thing you want to make sure the robot can do is drive across that terrain without falling into a crevasse or over a small cliff. By mounting the sensor pointing downward, you can achieve this quite easily.

Another common use of the ultrasonic sensor in classroom robotics has to do with moving objects. By turning the robot back and forth, or by mounting the ultrasonic sensor on a motor and moving it back and forth, you can create a sort of radar for your robot. If you want to increase the radar effect, you can add a second **axis** of rotation for up and down movement. You could practice this by having your robot compete in a battle where it has to strike enemies before they strike. Being able to sense your enemies before they sense you is key to winning the day. And when it's a battle you're discussing, the ultrasonic sensor can be of great use in a Sumobot battle.

TABLETOP TIP

A good way to practice in school for driving over uneven terrain with your robot is to place it on a table and have it drive forward. You'll want it to stop before it falls off the edge of the table. You can do this with the ultrasonic sensor. Instead of looking for something that is close, you'll be looking for something that is far away from you, or at least greater than the distance to the table.

In Sumobot, two robots oppose each other in a small circular ring, trying to knock each other off or out of commission. The ring is black with a white line around the perimeter. All robots are required to start in the center and "tag" the white perimeter. Once they have done that, it's a free-for-all. The simplest strategy is to drive straight forward and have the robots engage in a pushing match. But the smart robots, those built with cunning and programmed with precision, will slink around the outside, waiting until the ultrasonic sensor "sees" the opponent in front of it, and then strike quickly, pushing that robot from the side instead of head on and shoving it out of the ring. Go ahead and try it. You know you want to.

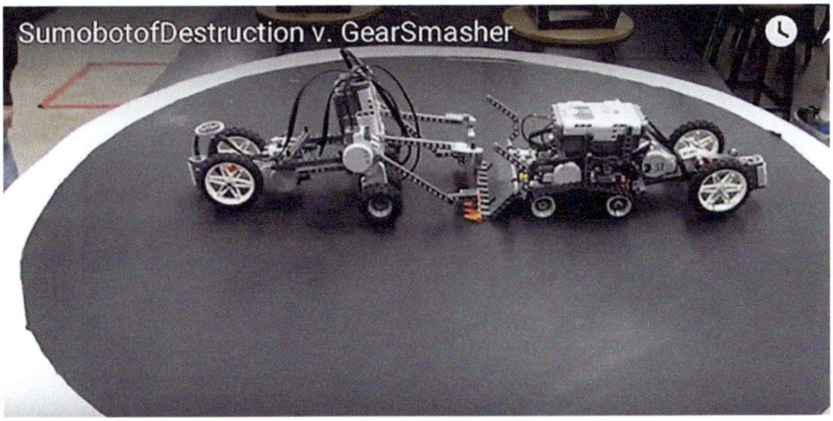

One of these fearsome looking Sumobots could gain the upper hand by using an ultrasonic sensor and sneaking up on its opponent.

Gyro Sensor

They gyro sensor is perhaps one of the least used and most misunderstood of the sensors on educational robots. A lot of students want to use the gyro sensor to make their robot stop after turning a certain number of degrees. One of the basic tutorials in the EV3 software shows the student how to do just this. Unfortunately, it rarely works, and some people just give up on the gyro sensor. This isn't fair to the gyro sensor as it can be quite useful when applied correctly.

Everyone understands light and color, which make the next sensors much easier to write about. But the gyro sensor is a little more finicky. According to LEGO, it detects **rotational motion**. Others say it measures **angular velocity** (the velocity of a body rotating around a specific axis; it is measured as the rate of change in an angle with respect to time). I'll let physicists argue about what the gyro sensor is measuring and what information it is actually displaying. For our purposes, it's important to understand that neither of these two terms is exactly the same as saying, "The robot has turned this particular angle." Close, but not quite. As I stated, that is what a lot of people are led to believe it does.

So what can it do? The gyro senor is very good at detecting circular movement around a singular axis. Therefore, if you want to know if your robot has tipped over, spun upside down, fallen off a cliff, etc., the gyro sensor can do that. In other words, if oriented correctly, the gyro sensor can tell if your robot has tilted up or down. Here is a good use: Build robots using gear reductions to try to drive up the steepest hill possible. Rather than only have set designations of steepness for the robots to attempt, let's say 30, 45, 60, and 75 degrees, it would be great to use the gyro sensor to record the robot driving up an ever steepening path. Whenever the robot's forward motion stops, you can read the gyro's output to see exactly what angle the robot had reached before it stopped climbing. Instead of my preset intervals of 15 degrees, you could differentiate between two robots that drove up 68.9 and 71.2 degrees.

Another nifty use of the gyro sensor was suggested by LEGO in a module from their Engineering and Design addition to the EV3 software. Students built a ramp to test the **coefficient of friction** (the ratio of the force necessary to move one surface horizontally over another and the pressure between the two surfaces; the lower the ratio, the less friction between objects—such as a hockey puck traveling on ice) of certain objects: a beam, a tire, etc. The gyro sensor is attached parallel to the ramp, and an object is placed on the ramp. As the ramp gets steeper, the user can press a touch sensor when the object starts to fall to record the angle at which it slid off. This can be used as a comparison to the angle at which other objects slide off. The gyro sensor will accurately record the angle at which the ramp stopped. It can also be used to test the **viscosity** of different liquids by spreading a piece of plastic wrap over the ramp and having the gyro sensor record the angle at which each liquid started to flow.

Don't think that gyro sensors are useless. When used to do what they are designed for, they work quite well. It is possible to use one to stop a robot turning at a specific angle—it just takes a lot of adjustments, which we will learn about later. For now, it's most important that you understand it is designed best to read angular velocity, not keep up with a moving robot and stop it on a dime while turning.

Color/Light Sensor

Some color/light sensors are designed to track lines using the reflection of relative light and dark, while some are designed to read colors, and others do both of these and then some. Most of the light sensor varieties work in two modes: ambient and reflective.

Ambient light mode is often an overlooked way of using a light sensor, but it's actually quite useful. Let's say you want to build an owl, and through your study you've discovered they are nocturnal. So you want to make your animal open its eyes when it gets dark out. For this, you would use the light sensor in ambient mode. Ambient light refers to the amount of light in an area. To understand this, close your eyes and look up at the overhead lights. Pass your hand slowly back and forth across your closed eyes. The general change in darkness and light that you experience when doing this is ambient light. It's the same

A light sensor uses one bulb to send out light and the second bulb to measure how much of the light is reflected back.

change in light that helps the owl know it's time to wake up and start hunting mice.

You can use a light sensor in ambient mode to monitor how many hours of daylight are in a given day or in conjunction with a sound sensor to determine how much noise and light are connected in a setting, such as outside your bedroom window.

Most light sensors are used in their **reflective light** mode. If you look carefully at a light sensor such as the one shown here, you will see two bulbs. One bulb shoots a beam of light at whatever surface you're pointing it at, and the other bulb reads what percentage of the light was reflected back.

The way a light sensor detects color is easy to understand when compared to how the gyro sensor works. Ask yourself why your mom told you to wear light colors in the hot summer. I'll bet you remember it's because light colors reflect more of the sun's rays than dark colors do. And it's the same with colors and the beams of light from the sensor. A dark color will absorb more of the beam than a light color will. Therefore, the reading for a dark color will be lower, meaning more light was absorbed and less was reflected back.

This field from the VEX Robotics Competition uses dark-colored lines on a white background to help robots find their way.

This leads us to a great many possibilities and uses of the light sensor. The most common is to program your robot to stop on a dark line or light line. The tables in my classroom are painted in whiteboard paint, so it's convenient to draw black lines on them. The difference between the percentage of light reflected back from the tables and the lines is large. This usually makes this an easy exercise to start with, but the difficulty can be increased by asking students to stop their robot on the third or fourth in a series of lines.

It is not by coincidence or happenstance that the playing fields for most popular robotics competitions include lines that are in stark contrast to the brightness of the areas around them.

Following a Line

One of the most awe-inducing and just plain neat things you can do to catch somebody's eye with your robot is to teach it to **line follow**: not to just drive and stop when it senses a dark line but to actually follow that line around a surface. There are many different ways to do this, from zigzagging back and forth along the edge of a line to following in a more straightforward manner using two light sensors to the ultra cool and smooth PID line follower—that's **proportional-integral-derivative**, and that's for the advanced programming at the end of the book.

Line following offers a lot of possibilities for our robot. Consider again a self-driving car and how using the lines painted on the pavement of highways and expressways can offer just another level of safety and guidance, provided the sensors are used accurately.

At this restaurant, the robot waiters follow white lines on the floor that contrast with a black stripe to deliver food to your table.

There are color sensors that are altogether different from light sensors, though they appear similar because they are built with light bulbs that shoot beams of light. The first thing to understand about your color sensors is their limitations. The LEGO EV3 sensors will work most reliably with six specific colors (red, blue, green, yellow, white, and black) and will work best if these are LEGO colors. This means it will respond better to a yellow piece of LEGO than it will to a yellow color card from the paint section of your local hardware store. The VEX color sensor works best with red, blue, and green but has a separate mode that allows it to expand to twelve colors. It also has an object near/not near setting.

Despite these differences, the basics are the same: to detect and react to colors. The uses of a color sensor can really only be limited by your imagination. You can create a simple color code so that your robot performs different actions based on which color it sees. Red for stop and green for go are good beginner challenges. You can program an animal that only eats food of a certain color, or you can make a search and rescue robot avoid certain colors on the floor that represent molten lava or rough terrain. Earlier I referenced a candy sorter. You can use the color sensor to sort your favorite candies, or even LEGO pieces themselves. The possibilities with the color sensor are endless.

Sound Sensor

It is a mystery as to why LEGO Education discontinued the sound sensor and as to why more educational robotics kits don't include one. The LEGO NXT sound sensor was a mainstay of the NXT kit but didn't make the cut with the upgrade to EV3. However, the EV3 software does have an option for the sound sensor, highlighting its importance.

The sound sensor converts decibels into a scale of 1–100. The maximum noise that it will register is about a loud yell, which stands out in a classroom but is a lot less loud than a jet plane. But since there are few jet planes or rock concerts in classrooms, there are some good uses for the sound sensor. One way in which it can be used is to measure the amount of noise in a room. Students can set up a program to record the noise level during a class period and then graph the results. Or, a teacher can place a robot at the front of the room and have it raise a "QUIET" flag when the noise reaches a certain disruptive level.

Another use of the sound sensor is similar to the touch sensor, in that it can be used to start and stop a program. I love to watch my students yelling at their robots to make them move around the room. Even funnier is to program the robots to move when they hear a loud noise and to stop when it's quiet. When this program is repeated, students keep up a loud cacophony until they are hoarse or their robot runs out of power.

The number one way in which I like to use the sound sensor is to start programs. Imagine this: you have fifteen robots lined up in a row, all ready to execute a series of movements. It's too crowded for students to kneel next to each other, and even if they do, they jolt the robots out of alignment by pushing the start button. On top of this, with two students to a robot you have thirty loud teenagers that make it difficult to hear the teacher's instructions. But the use of the sound sensor changes

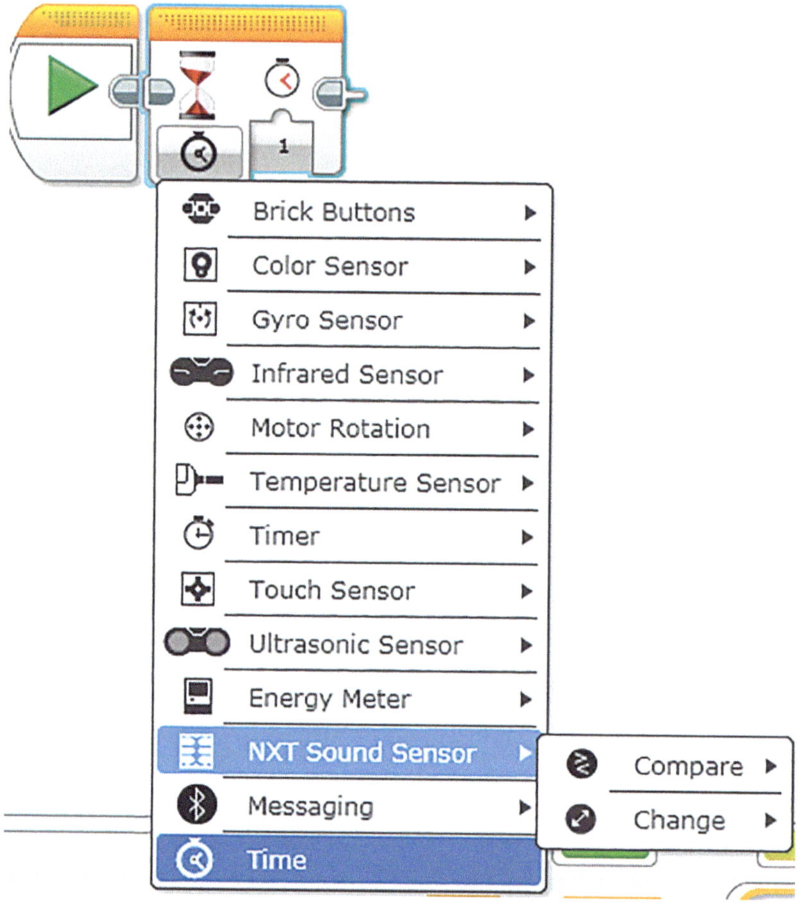

It was so popular, the only NXT sensor that works with the EV3 language is the sound sensor.

this. Students line up the robots then back away. Instead of talking loudly, they are waiting in silent anticipation of my fingers counting down, "Three, two, one," so they can all yell, "Go!" and set their robots on their way. The best part is if you have one goofball in the class who won't keep quiet, it's the other students who get angry at him/her for ruining the program and the teacher doesn't have to lift a finger.

These are the types of sensors used in competitions and a few of their most common uses. Students should let their imaginations run wild with ideas in how they would use them. Once they have some ideas bubbling their brain, they can use the next chapters to learn how to build with and place these sensors for maximum effect and how to program with them.

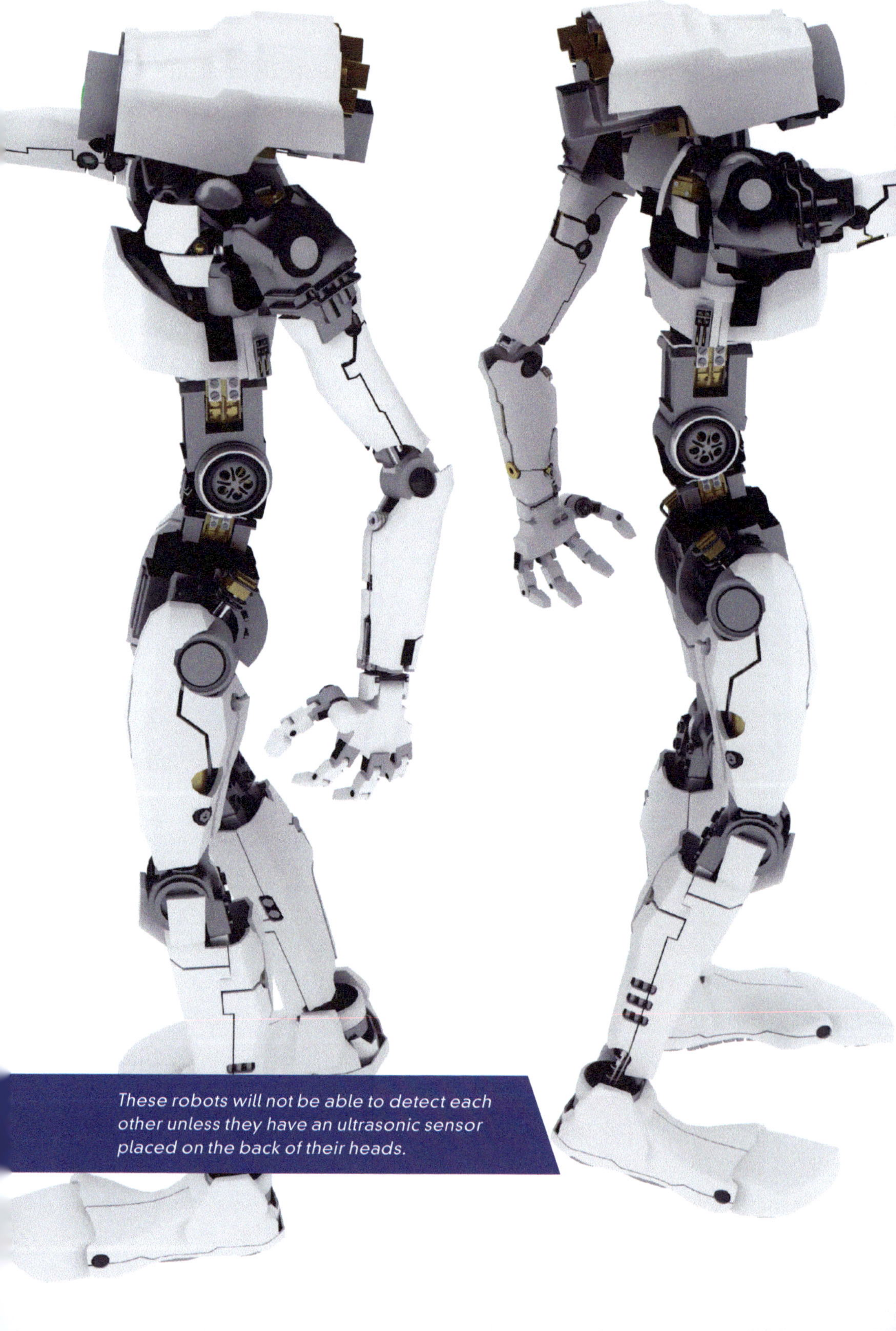

These robots will not be able to detect each other unless they have an ultrasonic sensor placed on the back of their heads.

3 Attaching Your Sensors

It's like you have eyes in the back of your head." While usually extended toward somebody as a compliment, the only time this really is a compliment is when you need eyes in the back of your head—otherwise they're pretty useless and creepy looking. At the beginning of the first two chapters, we talked about walking into a wall. If you're walking forward then there's no way eyes in the back of your head are going to do you any good. Put simply, you need your eyes positioned where they're going to do you the most good: front or back is all relative to what's going on around you. And it's the same with sensors on your robot.

This chapter will cover where to place sensors for maximum efficiency. Sensors are only as good as the information they receive, and they will only give you good information if you place them correctly. Think about our self-driving car again. Remember those sixty-four laser beams mounted on top? To give the car accurate information about everything in all directions, you need to mount them so they cover 360 degrees.

This car uses evenly spaced sensors to create a 3-D image of its surroundings, which will provide it with the necessary information.

They will have to be separated evenly, in this case with a distance of 5.625 degrees between each of them. You also have to make sure that they're not going to come loose while the car is driving over potholes, uneven terrain, or in inclement weather. And if some are pointing up while others are pointing down, then the car is going to receive less than desirable information. Similar considerations are going to apply to placing sensors on our robot.

One consideration we should look at with all sensors is how they will connect to the brain of our robots. In most pictures of robotics sensors, wires are not shown. This is because they can obscure the sensor. However, all sensors connect to the robot brain with wires, so it's important to understand two things.

The first is to use the shortest wire that does the job without stressing or pulling any parts. The second is to know your ports. The EV3 has four specific sensor ports on one side of the brain, numbered one through four, and four motor ports on the other side, lettered A through D.

The VEX IQ has six ports on each side that can accept sensors or motors. This makes placement of the sensors a little more important when it comes to the EV3. Other robots may have other configurations; it's important to know each of them before you start building.

It's up to you—the budding engineer—to figure out the best placement of each sensor in each situation. This book will provide some basic "getting started" information. It's up to you to take it and apply it to the great robotic creations you build. Let's get started by looking at how to build with the touch sensor.

Touch Sensor

When we ask, "Where is the best place to put the touch sensor?" the question we really should be asking is, "What do you want to use the touch sensor for?" Understanding that placement is almost always determined by intended use is really fundamental. Form will follow function, and this theme will be repeated as we learn to connect our sensors to our robots.

There are two basic ways the touch sensor is usually used. One is when you want a button to press to initiate part of a program. The second is when you want to use it as a bumper to detect touches, either to align your robot or avoid obstacles. The first of these two situations is the easier one to deal with.

TIP ON ATTACHING SENSOR

Most of the LEGO EV3 sensors will attach easily with the use of a gray double pin connector.

Students have an unnerving habit of connecting sensors using only one pin. This causes them to swivel and makes them essentially useless because they are no longer oriented in the desired direction. Using two connector pins will ensure this does not happen.

In the example below, the touch sensor is not being used in conjunction with the vehicle doing anything; it is just attached facing up so that it can be used to record a touch. For example, you may be using it to record how many times an event occurs by pushing it or using it to start and stop some sequence of programming on your robot. The touch sensor in this case can

The touch sensor is attached to the side of the robot using a double pin connector so it won't move.

easily work around your design demands, and you don't need to give it too much thought when building.

The second situation, in which you are using your touch sensor to detect the area around you, needs a bit more planning to make it work right. A touch sensor on its own is not ideal for stopping you from driving into an obstacle at full speed. Take a good close look at your touch sensor. No matter what type it is, it has a spring connected to a plastic surface that will be the initial point of contact in its native form.

In the case of the EV3 touch sensor, it's a red protrusion with an axle hole in the middle. The VEX IQ bump sensor has five pinholes with which to attach beams of various lengths and shapes. By themselves, they are effective only with a direct contact, like the push of a thumb, but if you want to use them effectively to align your robot next to an object or stop it from

The range of this VEX touch sensor may have been extended too far. Contact with the ends of the gray beam may not register as a touch.

crashing into things, you need to take the native form as a starting point and expand from there.

When the range of the touch sensor is extended beyond its native state, it gains the advantage of being able to detect an object that is somewhere in front of you as opposed to directly in front of you. However, to accomplish this you must guard against a design in which contact will not push in the sensor button or knock your sensor out of alignment. This trade-off becomes the balancing act with all touch sensors: extending their range to make them more effective, but not so far that they don't work. The photo on the previous page shows a good example of adding too much to a touch sensor. The beam attached to the touch sensor is probably too wide. It could pull the robot sideways if hit on the end rather than register a touch.

TOUCH AND GO

To finish our look at the touch sensor, remember these three things:

1. Be clear about what you want to use the touch sensor for, then build for that specific situation.
2. A touch sensor is not good for stopping a fast car from crashing.
3. Extend the range of the touch sensor by adding pieces, but don't overdo it.

Ultrasonic Sensor

When used correctly, the ultrasonic sensor will not be bumped into nearly as many times as the touch sensor. So our main concern with the ultrasonic sensor will be in placing it for the most efficient readings, rather than to absorb maximum impact safely. Just as we did with the touch sensor, we want to ask ourselves what we are going to use the ultrasonic sensor for. Is it to register the passage of an item in an assembly line? Is it to search out objects in front of us, or to look for dangerous drops below the robot? Or is it to help us build and program a robot to parallel park?

In each of these situations, you are going to place the ultrasonic or distance sensor on a different part of the robot or facing a different direction, but you are always going to follow the same rules when placing the ultrasonic sensor.

The ultrasonic sensor on this parallel parking robot is directly facing the books it is sensing. When the sensor no longer detects the books, the robot will change direction as programmed.

First and foremost, the ultrasonic sensor works best when it (or technically the high-frequency beam it shoots out) is perpendicular to the surface it is hitting. The easiest way to do this is to make sure the robot you are going to attach the ultrasonic sensor to is parallel to the ground. Specifically, the entire robot does not need to be parallel, but the ultrasonic sensor itself does, and this is most easily accomplished by using a robot that is not built on an angle, such as a race-car with its front lower than its back. Once you have a robot that rides parallel to the ground, attaching an ultrasonic sensor in the right configuration is easy.

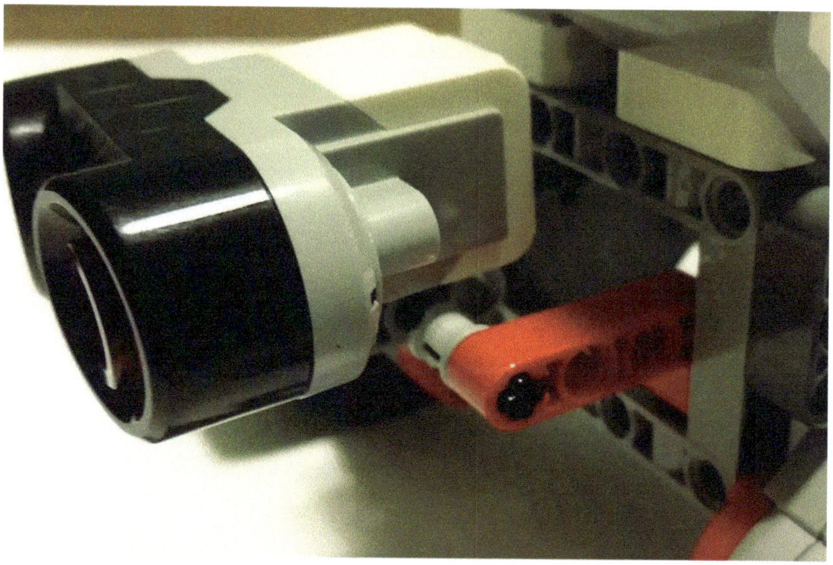

This ultrasonic sensor is parallel to the ground. This will enable the high-frequency waves to bounce directly back to it.

Damien Kee's Riley Rover is an example of a robot that meets these requirements; it has a very simple, quick, and effective design for classroom use. By simply using a right angle beam, a short axle, two pins, and a bushing, you can attach an ultrasonic sensor by taking advantage of the fact that all EV3 sensors have an axle hole as their middle attachment point. On the Riley Rover, the ultrasonic sensor seems to be staring straight ahead, ready to bounce a high-frequency beam directly off anything in its way.

Once you have a flat (parallel to the ground) robot, then you just need to know whether you are trying to sense an item in front of you, below you, to the side of you, etc., and build accordingly. The sensors should be placed parallel or perpendicular to your robot. If you place the sensor at an angle, it will not be as effective as the other configurations simply because the beam will not bounce directly back toward the sensor. However, this is a possible configuration when you set the sensor to look for objects at a distance. If it spots objects that are farther away than the surface beneath it, that is an indication the edge of the surface is close by.

So far we've dealt with keeping the ultrasonic sensor pointed perpendicular to the item it's looking for and making

WRONG SIGNAL

One consideration that affects ultrasonic sensors more than any others is wires. Wires usually won't move a sensor or make noise to set off a sound sensor, but their general proximity can do funny things to an ultrasonic sensor. Take care to route the wires behind the sensor and away from anywhere they might cause an errant reading. A wire doesn't have to be directly in front of the "eyes" on the distance sensor to give you an inaccurate reading and drive you crazy.

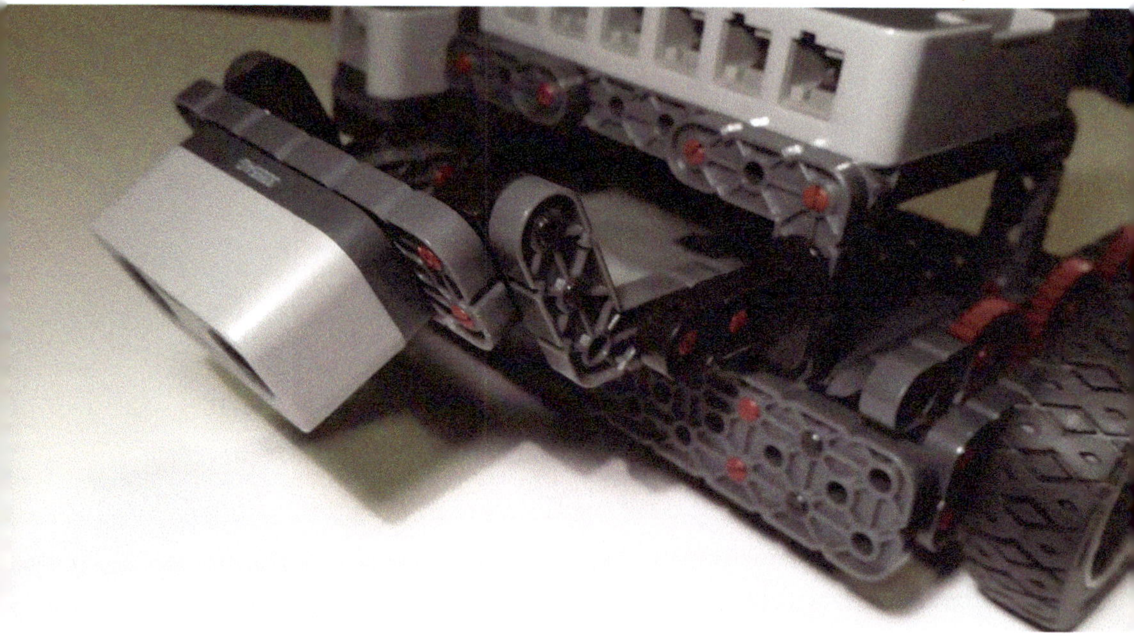

This ultrasonic sensor is mounted on an angle. It may not accurately return distance readings as the waves may bounce out at an angle.

sure the wires are not in the way. The third consideration for ultrasonic sensors doesn't deal with them directly, but with the surface they are trying to sense. Ultrasonic sensors will work better when they reflect off a flat, hard surface. This has to do with that high-frequency beam bouncing directly back to the sensor and providing an accurate reading. If you drive toward a cereal box and a soda bottle, you're going to have an easier time sensing the cereal box. And if you're trying to keep the robots on a table by having the ultrasonic sensors facing down, you may have a bit better luck with a hard floor rather than a carpeted surface. These are the ideal conditions, but the world is not ideal, so it's OK to challenge yourself in less than ideal conditions. Just be aware of the possible issues you may face when using the ultrasonic sensor.

Gyro Sensor

There are fewer issues to attend to when attaching a gyro sensor to your robot. The main point you have to consider is this: In which direction do you want to detect a change in angle—or rotational motion, to be precise? The gyro sensor (at least the commercial ones available in most kits produced for the education market) can detect motion on only one axis, either *x*, *y*, or *z*.

To help understand this, keep your arm against your side and hold your hand out in front of you so it is parallel to the floor, then perform these three tasks: First, rotate your hand to the left and to the right while keeping it parallel to the floor. Second, rotate your wrists so the thumb moves from the side of your hand to the top of your hand. Third, flex your wrist so all your fingers point first up and then down, as if dribbling a basketball. These tasks mimic the three axes of motion. Our gyro sensors can detect rotational motion along any one of these axes, but not more than one at a time.

So you need to know in which direction you want to detect rotational motion: up and down, left to right, or side to side. The most common would probably be the left-to-right turning of your robot, or the first movement. Most students would use this situation to program their robots to make accurate turns. By sensing the rate of rotation across an axis, the robot can calculate the angle of the turn, but as LEGO Education says in the EV3 User Guide, "… with a +/– 3 degree accuracy." Experiments in my class seem to bear this out.

If you want to make a 90-degree turn or turn toward a certain feature in a competition setting, the gyro sensor can help you do this if you orient it correctly.

The other two **orientations** are less common but still important to understand. The second movement, where you twisted your wrist so your palm moved clockwise, might be similar to detecting how much rocking a boat does in a storm on the open sea. You may want to detect if your robot is tilting, such as it would when falling over a cliff. In this case it is like when your fingertips pointed forward but your hand moved up and down. For each of these three situations, it is easy to decide how to orient and attach the gyro sensor: just follow the arrows.

While you don't have to worry about wires getting in the way or protecting it from a crash, there are two final considerations for the gyro sensor. The first is to keep it from being tilted. If you look at the three examples on the next page, whichever direction the gyro sensor is detecting movement in, it's lying at a zero degree angle in reference to that movement. This is not as complex as it sounds.

For left to right movement the sensor is laying parallel to the ground (0 degrees) so that only left and right movement (as shown by the arrows) will be detected. The second you lift one end of the sensor and place it on an angle below or above 0, you're going to get some funky readings. The same is true with the other two orientations.

Finally, your gyro sensor will suffer from **drift**. This means that you will see sensor readings that change at a fast rate when the robot is absolutely still. It may drive you slightly crazy at first. These readings are normal for a sensor that costs less than

SENSE OF DIRECTION

The arrows show the direction of rotation the gyro sensor is set up to detect. By placing the gyro sensor the proper way, you can detect the motion you want.

This is the way you want to orient the arrows if you want to detect a turn to the left or right.

This is the way to orient the arrows if you want your gyro sensor to detect motion like the rocking of a boat from side to side.

This is the way to orient the arrows if you want to detect an up/down tilt, which would sense if the robot is falling off a cliff or climbing a big hill.

$100 as opposed to one that costs thousands, which apparently doesn't suffer from drift.

Damien Kee, an Australian roboticist, wrote in his book *Classroom Activities for the Busy Teacher: VEX IQ with Modkit for VEX*: "The Gyro can be extremely sensitive. It is important that the Gyro is extremely still whenever it is plugged in. When using the Gyro for the first time with a program, it is always handy to plug it in, place it on a stable surface so it is still and then turn the VEX IQ Robot Brain on. If you continue to run into problems with the Gyro, sometimes it helps to unplug and re-plug the sensor back into the VEX IQ Robot Brain to reset it."

Kee's comments apply equally as well to the EV3 gyro sensor. However, if you are using the EV3 iPad app, there is a great work-around. With just a tap of a single **icon** you can reset your gyro sensor to zero. Doing this once or twice tends to stop the drift.

Color/Light Sensor

In order to understand how to build and use our color and light sensors, it's important to remember the different modes in which they operate and to go back and look at the history and development of these sensors. The color and light sensors available with most educational robotics kits have a number of similar but not identical functions. Color sensors have come a long way in a relatively short period of time.

When LEGO had its NXT line, it produced a light sensor that had reflective and ambient light modes. VEX had a sensor for its EDR line which they called a **line tracker** presumably because that was its intended use, even though it functioned the same way as a light sensor in reflective mode. For a while, to find a true color sensor for your NXT robot you had to go outside of the LEGO Mindstorms products to a third-party company like HiTechnic to find a color sensor compatible with the NXT.

All that has changed now, and both VEX and LEGO offer combo sensors that can perform different functions. LEGO's color/light sensor has three modes of operation: ambient light, reflective light, and color. And VEX offers a sensor that senses reflective light and color (though different colors than LEGO) as well.

When working in ambient light mode, you are looking for the given amount of light in an area. This is what you are looking for if you are trying to determine if it's day or night outside. This might be very useful for the self-driving car because, regardless of what time the car's computer says it is, the amount of light outside might affect a whole host of other readings. So in the case of using ambient light mode, the only consideration when building is to make sure the sensor is free from obstructions and pointing in the general direction of where you want to get a light reading. In other words, don't point it at the ground or bury it underneath the robot. Anywhere on top pointing up is fine.

When using these sensors in the other two modes, reflective light or color, similar rules apply for building. The first is that you want to be close to, but not on top of, the surface from which you are trying to get a reading. If you're too close, your sensor won't read anything but black or really dark, and if your sensor is too far away, bright surfaces won't reflect properly and your colors will blend and be indistinguishable from each other. Approximately 0.5 inches (1.25 cm) above or away from the surface you are trying to read is usually ideal.

STAYING THE COURSE

Here's an example of why it's crucial to properly place a light sensor. For a drag race down a 20-foot (6 m) by 4-foot (1.2 m) bright yellow board, black electrical tape marks the finish line.

A light sensor is affixed to the bottom of the drag racers to sense the reflection of the black line, which is obviously darker than the yellow. The robots will drive forward fast on the yellow and stop on black, and also display the elapsed time (this way many cars can race without being on the track at the same time). If the sensors are too high off the ground, they would read cracks or dirt as the black tape and stop too early.

Usually, about 0.5 inches (1.25 cm) clearance between the sensor and the surface beneath works best.

As was the case with the ultrasonic sensor, sending out a straight beam of light that can bounce back off of a perpendicular surface will improve your chance of getting an accurate sensor reading.

Once you have your sensor positioned so that it is about 0.5 inches away from whatever it is going to read, the second consideration is protecting it. In this case, you are not trying to protect it from an impact, which the touch sensors are liable to take. Instead, you are protecting it from outside light sources. Whether we are reading color or reflected light, any outside light source (shadows, bright sunlight reflecting off of a pair of glasses, cellphones snapping pictures with the flash on, overhead fluorescents, etc.) is going to interfere with our readings and not allow us to obtain accurate information.

Rather than be told exactly how to attach a sensor, you should work with what you have to apply the concepts espoused here. You don't need to use the exact pieces used on our example robot. Use your own ingenuity to come up with something similar.

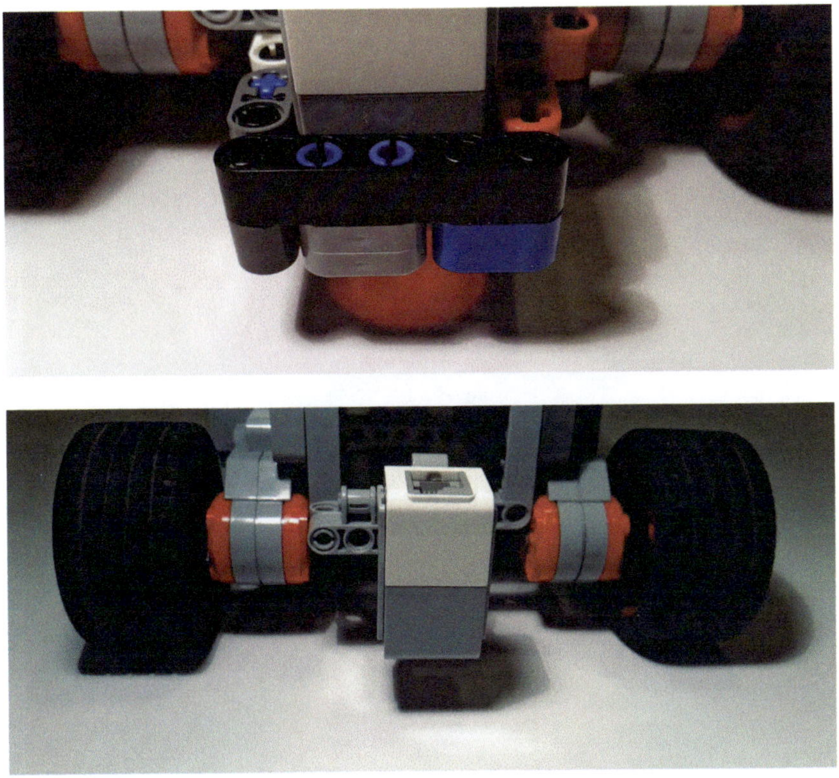

The top image shows a light sensor with shielding to protect it from outside light sources. The bottom images shows a sensor without a shield.

There is one more problem to address with light sensors: shadows. The purpose of a color sensor shield is to protect the color sensor from shadows that may be cast by a person (team member, student) or object (like a competition module), and to protect the beam of light from spreading and returning readings from a wider area than you intended to examine. The more focused your beam is, the more accurate your readings will be. There will always be some shadow cast by the casing of the color sensor, but the beam of light that shoots out and hits the target is usually bright enough to counter the effect of this.

Other Sensors

We have covered the most prevalent sensors that you will find in most educational robotics kits. However, there are many others out there. VEX EDR has a special sensor called a

The LEGO Mindstorms temperature sensor detects the temperature of close objects. Don't allow your robot to get too close to hot objects.

potentiometer. You place a shaft through it and it will read how much the shaft turns. They are very useful if you want to stop something turning, such as a swivel bridge, but don't want the potential damage a touch sensor could undergo when it hits a wall each time the bridge reaches its mark.

The EV3 home edition, which is a very similar kit to the educational one, has an infrared sensor that looks a lot like the ultrasonic sensor. And while it is shaped the same as the ultrasonic sensor, remember, form follows function. The **infrared beacon** functions in a few different modes. In proximity mode, it uses light waves reflected off of an object to detect its proximity. When using this mode, it would be proper to attach it in a similar manner as we attached our ultrasonic sensor.

The infrared sensor also has a beacon mode that detects IR (**infrared**) signals sent from a special IR beacon (which also comes with the home edition) on one of four channels.

The sensor will relay two pieces of information to the robot based on what it senses: proximity of signal and general **heading**. In order to make best use of the general heading information, it is important to attach the infrared sensor in the center and front of your robot if that is the direction you are intending it to drive. But the requirements to be perpendicular to the object you are trying to detect aren't as strict as they are with the ultrasonic sensor. The beacon is broadcasting an IR signal in many directions and your infrared sensor will detect these. It doesn't have to be pointed directly at an object to detect high-frequency waves bouncing back as with the ultrasonic sensor.

KNOW THEIR LIMITATIONS

Before you program with all of these sensors, keep in mind that for any program to work, you have to understand the limitations of your sensors and attach them in a way that minimizes their weaknesses and maximizes their efficiency.

LEGO Education also makes a temperature sensor that has a long metal pole that extends from it. Again, form follows function. This is not a weather forecasting sensor for room temperature; it's designed to detect the temperature of close materials. One use would be to have a robot ring an alarm when a pot of water has boiled. You will need to build accordingly. In this example you don't want to burn your robot, so you will need to build an arm that extends away from the robot and up and over the pot of water so the temperature sensor can do its job.

Alternative Sources

As mentioned in the first chapter, this book focuses on VEX and LEGO as they are the most prevalent educational robotics systems out there. You will find some of the other options listed at the end of this book. Being the biggest and oldest company has not only allowed LEGO to have a full complement of sensors to equal that of VEX, but there are also many third-party sensors out there that are built to be compatible with the EV3 platform.

The best way to build with all of them is to remember that form follows function. You need to decide what you want to use the sensor for and then decide the best way to attach that sensor to your robot. This knowledge can be applied to any and all sensors out there.

Here's a quick sample of some of the third-party sensors available for EV3: HiTechnic makes sensors that look a lot like the EV3 sensors. Among their most popular ones are compass, magnet, **barometric**, infrared**, acceleration**, and **proximity sensors.** Vernier, a company that makes a lot of science equipment, has a large line of investigatory sensors available for EV3. In their list of more than twenty different sensors compatible with the EV3 are an **anemometer, current probe, pH sensor, salinity sensor, soil moisture sensor, UVA sensor, turbidity sensor**, and much more. Turbidity is the level of clarity of a cloudy or muddy liquid; an anemometer measures the speed of the wind or other gas.

Another popular third-party sensor company is MindSensors. Among the many sensors it makes is a range of cameras that can be attached to your robot to give it vision where it can see and detect objects, not just distances to or from them!

The motions of this pitcher would have to be programmed meticulously into a robot and would include hundreds of lines of code.

4 Commanding Your Robot

The nicest car, with the most advanced sensors, placed and attached in the correct fashion, still won't move an inch without software programmed to run it correctly. So far we've learned what sensors do, some peculiarities of the way they work, and how best to attach them to your robots. Now it is time to turn our attention to programming your robots to do what you want using your sensors.

Correctly stated, you are programming robots, not sensors, for the brain of the robot contains the processor that takes our program and executes it. Sensors themselves don't have any processing power (at least not the ones students are using). They only do what they're told to do by the robot brain. They receive commands from the brain to sense their environment, then return the values they sensed to the brain, and then the brain goes on to the next command. This is an important distinction as it helps us understand that sensors are part of a bigger system, and it's the overall system we'll be working on programming in this chapter.

A few notes on methodology here: As this book is for high school grades, the programming level will be presented somewhere between basic and intermediate. It is assumed readers have a basic level of understanding when it comes to using software. If all of this is new to you, try to follow along and ask questions of your robotics teammates responsible for programming.

To get started, we will try to follow the same pattern with the first few programs. We will start with the basic description of the task and then work on an algorithm to help us complete it. From there, I will offer pseudo-code to help us understand how to break the algorithm into a set of commands a robot could follow. Finally, I will write the robot code or provide screen shots of code for you. I will use examples in RobotC's 4.0 Graphical Language for the VEX IQ as well as in LEGO Mindstorms' EV3 programming. I will show one or two examples in RobotC, as this is a syntax language, meaning it is written in text rather than with graphical blocks and icons.

Tracking Your Steps

An algorithm is simply any set of steps used to solve a task. Teachers assign common ones to their computer science students all the time to get them to understand how to start thinking like a computer. For example, if you were asked to write down all the steps it takes to throw a ball to somebody, you may write something like this:

1. Pick up ball.
2. Raise arm overhead.
3. Swing arm in direction of friend.
4. Release ball when arm is on downward swing.

Your teacher may challenge you to include steps like this:

1. Look down and locate ball.
2. Reach down to pick up ball.
3. Grab ball with your throwing hand.
4. Stand back up.
5. Locate your friend and stand facing her.
6. Take one step forward with leg on non-throwing side of body, while at the same time …
7. Reach throwing arm down and low and start to bring it straight up behind you in a circular motion.
8. When your arm is on a line with leg on non-throwing side, open your fingers so as to release the ball.
9. Follow through with your arm in a downward motion.
10. Recover your balance.

Your teacher isn't being mean when making you double the number of steps you thought it took to do a task; he or she is simply trying to get you to see the world the way a computer does. So much of what we do as humans is natural and done without thought and is a combination of so many different actions that we barely think about them. But with a robot, we have to. Remember, a robot doesn't have experience, understand self-preservation, or think for itself. And yes, scientists are working on that, because the robot you use in school doesn't have artificial intelligence, yet.

When we write something in pseudo-code we are breaking our algorithm down into individual commands we want a robot to execute, though we are not using any computer-specific language at this point. Finally, once we have our pseudo-code, we then follow the rules of our programming language to translate it into commands a robot can follow.

Sensor Ports and Reading Sensor Values

Because this next topic will apply to all sensors in both programming platforms, I wanted to touch on it before we examine specific sensors. EV3 has an auto-ID function that recognizes what type of sensor you have plugged into each port. This usually allows accurate port numbers to display in your programming environment. In other words, if the touch sensor is plugged into port 2 and you connect your robot to your computer (all programs are downloaded from your computer into the brain of the robot), when you pull up a touch sensor icon, it will automatically default to port 2. In RobotC, you need to identify your sensors by clicking the Motor and Sensor Setup icon at the top of the screen. It will bring you to a page that looks like the image at the top of page 61.

You can name the sensors whatever you like, but the default names were kept in this image. One thing you should do is follow the syntax of most programming languages: use two

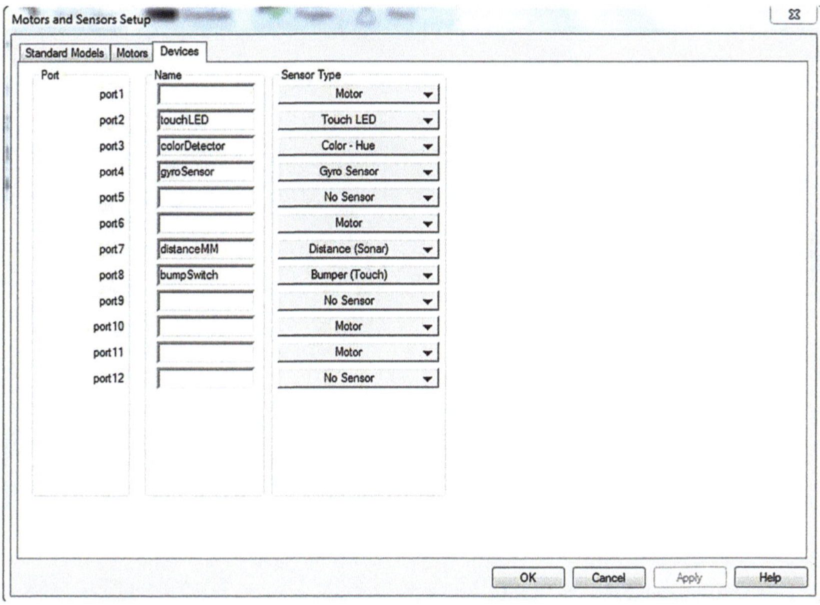

This is the Motors and Sensors setup page in RobotC. Take note of the nomenclature used in naming each piece.

words, no spaces, capitalize the first letter of the second word, but not the first.

Like this: touchSensor

Unlike EV3, you don't have to worry about which port you connect the sensor to as all twelve ports accept any sensor or motor on the VEX IQ.

Displaying the sensor values is also easy to do. We're going to use the "Port View" option on our EV3 robot to read this number. (The screen on an EV3 is not backlit so it can be hard to read clearly.) First, turn on your robot and make sure it has an ultrasonic sensor plugged in. From the home screen of your robot, navigate over to the third tab using the right arrow. This tab is called the Brick Apps and the top choice is called Port View. Once you've navigated over to Port View, you choose it,

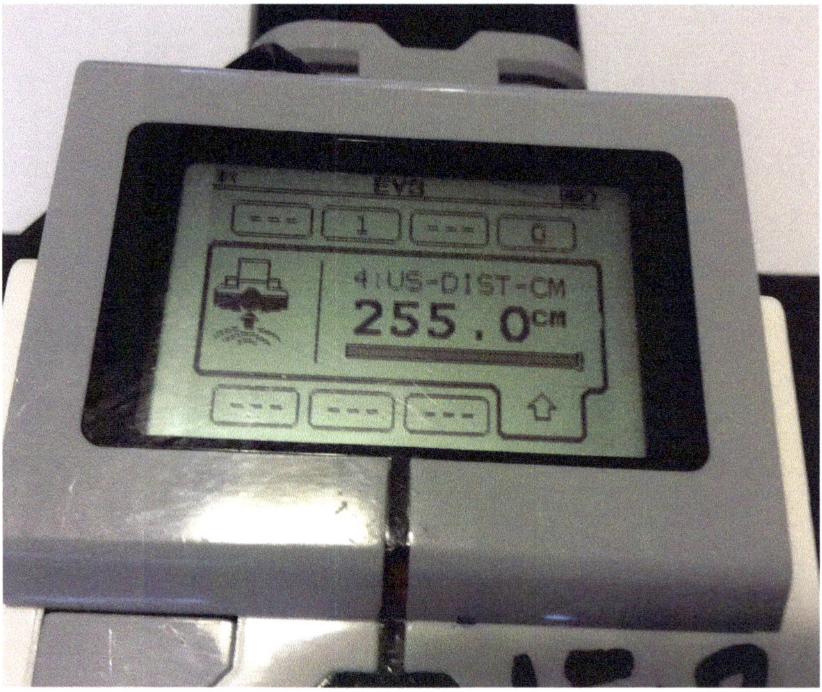

The Brick Apps is the third tab on the EV3 screen and contains the "Port View" tool, which we use to read our robot's sensors and motors.

and then you have to navigate toward the correct port. I chose port 4, which you can see in the image above. The reading in that image is when there was nothing in its way. The closest wall was at least 8.35 feet (2.6 m) away. It reads "255 cm" (100 inches), which is the maximum distance the ultrasonic sensor can read.

```
repeat (forever ) {
    displaySensorValues ( line1 , colorDetector );
}
```

This program will display the value of the color sensor you labeled "colorDetector" on line one of the VEX IQ LCD screen.

While the VEX IQ screen is easier to see, there is no Port View function to let you read the sensors. You have to write yourself a program like the one at the bottom of page 62.

Programming with the Touch Sensor

Let's pretend we have built a fun robot that is designed to throw a plastic ball at our friends as they run by. The robot will throw the ball when you push the touch sensor. The algorithm I would write would look like this:

1. Push the touch sensor.
2. Throw the ball.

That's pretty simple and straightforward. The pseudo-code for a program this simple is not much different, but it will include specific robot-related commands. So the pseudo-code would be written as:

1. Wait for the touch sensor to be pushed.
2. Turn on the motor attached to the arm at full power for 1 second.

This program says (from left) *when the program starts, wait for the touch sensor connected to port 1 to register a touch, then turn on the motor connected to port D at full power (100) for 1 second.*

FOUND IN TRANSLATION

There is a nifty feature in RobotC that allows you to convert a graphical program into a text-based program. This is really great for the beginning programmer so she or he can learn how the syntax works in more complex programs. The reverse does not work simply because there are too many features in RobotC that do not work in RobotC Graphical. To find this feature, click View>Convert Graphical File to Text. The images to the right show how the same "touch then throw" program is written in RobotC's graphical environment for VEX IQ and in the RobotC syntax language.

The syntax in a text-based program may seem confusing at first, but if you compare this to the graphical program above it, you should be able to decipher what is happening. And it's not as much writing as it seems. RobotC automatically generates all the "pragma" lines at the top after you fill out the Motor and Sensor Setup.

The programming block and the text block at right show two versions of the same program. Both will activate an arm for one second after a touch sensor is pushed. By viewing the two images you can see how the graphical blocks are represented by the syntax in the text-based code below.

```
waitUntil ( getBumperValue(bumpSwitch) == true
setMotor ( motor1 , 100 );
waitUntil ( getTimer(T1, seconds) == 1
stopMotor ( motor1 );
```

```
#pragma config(Sensor, port2, touchLED, sensorVexIQ_LED)
#pragma config(Sensor, port3, colorDetectorLED, sensorVexIQ_ColorHue)
#pragma config(Sensor, port4, gyroSensor, sensorVexIQ_Gyro)
#pragma config(Sensor, port7, distanceMM, sensorVexIQ_Distance)
#pragma config(Sensor, port8, bumpSwitch, sensorVexIQ_Touch)
#pragma config(Motor, motor1, leftMotor,tmotorVexIQ, openLoop, driveLeft,
    encoder)
#pragma config(Motor, motor6, rightMotor, tmotorVexIQ, openLoop,
    reversed, driveRight, encoder)
#pragma config(Motor, motor10, armMotor, tmotorVexIQ, openLoop,
    encoder)
#pragma config(Motor, motor11, clawMotor, tmotorVexIQ, openLoop,
    encoder)
//*!!Code automatically generated by 'ROBOTC configuration wizard !!*//

taskmain ()
{
    waitUntil (getBumperValue (bumpSwitch) == true) ;
    setMotor (motor1, 100) ;
    wait (1, seconds) ;
    stopMotor (motor1) ;
}
]
```

See the difference? As humans, we know what "throw the ball" means, but to a robot, it means a motor is going to have to move a throwing arm. Now let's look at how this simple program would be written in EV3:

The middle icon in the image below is called a "wait for" command and is quite popular for use with EV3. You will find the "wait for" commands in the orange colored "Flow" palette at the bottom of the EV3 screen. The "wait for" icon will not contain any values because in its default state it's set to wait for time. You have to change it to the sensor you wish to use by clicking the bottom left corner and choosing the correct parameters.

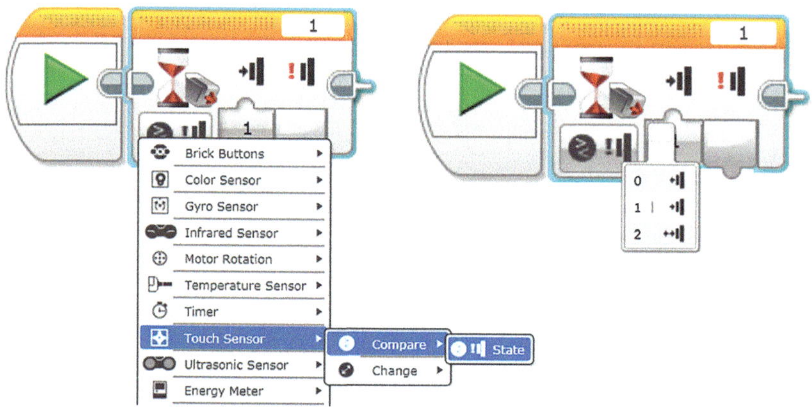

Wherever you want to change a value on an icon in EV3, you click in that area. The one place not clicked in these examples is the top right, where you can see the number 1. That refers to the port to which the sensor is connected.

In the above examle, you can see I chose the Touch Sensor>Compare>State. Similarly, I clicked the button directly to the right on the bottom of the icon to choose the state that I wanted, which in this case was option 1, pushed in.

When programming with sensors, always use this mantra: "Do something, sense something, do something else." This simple mantra is written on the board in my class and repeated by my students, but they often forget the last part, to do something else. For example, they will have their robot drive forward over a white surface and then sense a black line, but the robot doesn't stop. Usually it's because they forgot the last part, "Do something else," which in this case would be to stop the motor. Even though they still forget, having the mantra makes it easier to teach and catch mistakes.

READING YOUR ROBOT

Using Port View is a great way to learn how your sensors work. Plug in all of them, navigate to Port View, and walk around examining distances, colors, light reflections, angles, etc. After doing this for a while, you'll really get a good sense of what information your robot can and cannot receive from sensors.

In the programs shown, it may look like only two things are happening, the "sense something" and the "do something else." However, in these simplest of programs, the robot sitting still is covering the first part. This is the something it is first doing, and after the button is pushed, it does something else.

Ultrasonic Sensor

I am going to add a small level of complexity to the previous challenge. Instead of waiting for your friend to pass by so you can push the touch sensor, we're going to let the robot autonomously sense your friend by using the ultrasonic (distance) sensor. Here's the algorithm I would use for this situation:

1. Wait for friend to pass in front of robot.
2. Throw ball.

Again, we only have two actions, but the robot sitting still and waiting covers the "do something." The pseudo-code for this program would be:

1. Wait until the ultrasonic sensor senses a close object.
2. Turn on the motor attached to the arm at full power for one second.

EXTRASENSORY PERCEPTION

Bob Biamonte, a robotics educator, warns of a problem when using ultrasonic sensors. In an email from the LEGO Engineering Google Group, he wrote: "We tried a follow the wall challenge using the ultrasonic sensor and all the robots started doing strange things. Then it hit me that twelve EV3s using ultrasound in the same room were all hearing each other with great confusion." So, in settings with multiple robots all using their ultrasonic sensors at the same time, you may get some inaccurate readings.

Before we proceed with the program in EV3, we have to learn how to read our sensors. If you recall, the ultrasonic sensor works similarly to the echolocation of animals. It sends out a high-frequency signal and waits for that signal to bounce back. The rate of the bounce back helps it determine how far away an object is.

I'm going to set up my robot so the ultrasonic sensor is facing a wall that is more than 255 centimeters away, which means it should detect any object that moves in front of it closer than that distance. The program in EV3 therefore would look like this:

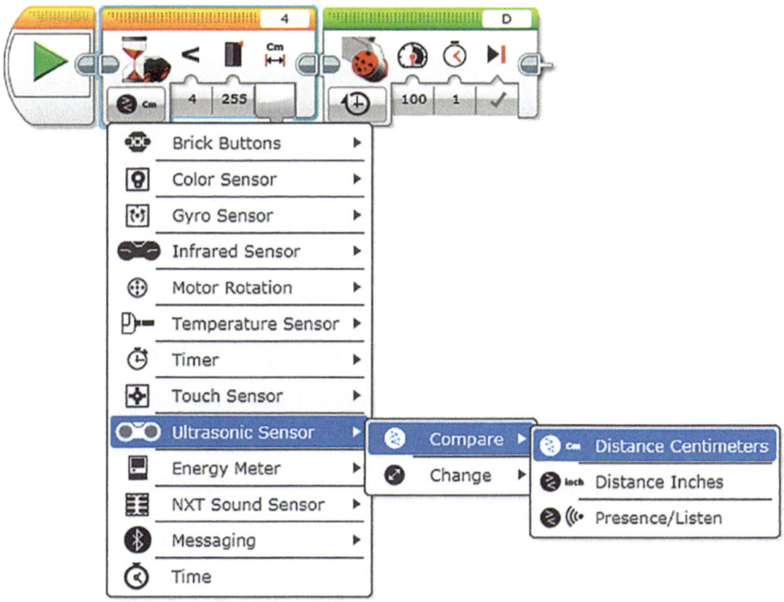

This is an image in which the options were left open so you can see that Ultrasonic Sensor>Compare>Distance Centimeters was chosen.

With the above program, when any object gets in front of the ultrasonic sensor at less than 255 centimeters, it will sense it and then the arm attached to motor D will move forward at power level 100 for 1 second. The numerical value under the “less than” symbol is the program’s designated number for selecting the less than option.

The ultrasonic sensor program that is probably taught first by most teachers is to drive toward an object and stop at a certain distance from that object. Let’s say 10 centimeters (4 inches) for sake of argument. Here’s the algorithm:

1. Drive forward.
2. When you see a wall 10 centimeters away, stop.

And here’s the pseudo-code:

1. Turn both drive motors on at a medium power.
2. Sense an object closer than 10 centimeters.
3. Turn both motors off.

Want to see the program? Try and figure it out for yourself. The answer will be in Example One on page 85 for both EV3 and RobotC Graphical for VEX IQ. Just remember, do something, sense something, do something else.

A great beginner challenge is to have students program their robots to continuously move as long as there is a loud noise level. This challenge will help us look at how to use loops, sometimes called “while” statements.

We could write the algorithm and pseudo-code like this:

Algorithm

1. If you hear a loud noise, move.
2. Keep moving until you stop hearing the loud noise.
3. Stop.
4. Go back to the beginning and listen for a loud noise.

Pseudo-code

1. Wait for sound sensor to hear loud noise.
2. When loud noise is heard turn motors on.
3. While sensor hears loud noise, keep motors on.
4. When loud noise is not heard, turn motors off.

Here's the program in EV3:

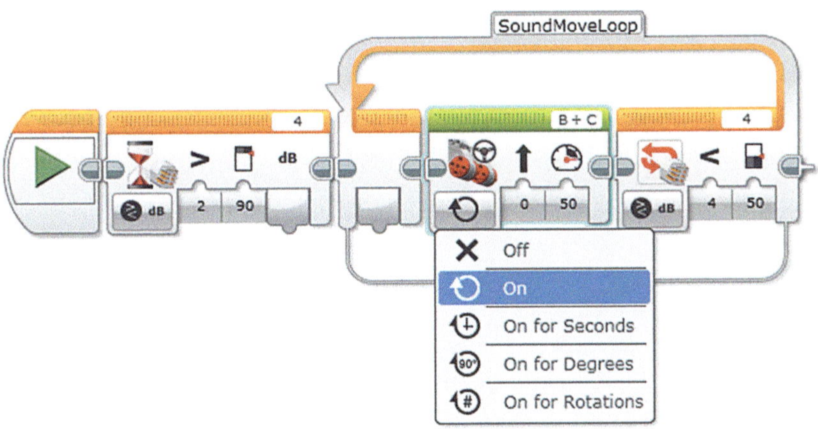

One of the keys to programming a robot to move until its environment changes is to turn the motors on, and not just for a designated period.

There's a lot to look at in this simple program. First is the use of the loop. In EV3, you use a loop whenever you want to repeat an action. If you've ever used Scratch or done some of the Hour of Code challenges, you may be used to loops that automatically encapsulate the code you want to repeat. In EV3, you have to drag a loop out, then highlight your code you want to repeat and drag it inside the loop. In RobotC, repeating an action can be accomplished by using a "while" or "repeat" function. Both do the same thing. They repeat an action or set of actions until told to stop. In this case, I've set the loop to stop when the sound sensor senses noise lower than 50.

TIME TIPS

You can turn motors on for different durations: time, rotations, or seconds. Or you can just turn them on. If you just turn them on, there is nothing internal telling them to stop. They will keep going until something else happens. This is the setting you want for your motors whenever you want them to run until a sensor senses something. Not using this setting is a common mistake students make.

(The sound sensor reads decibels in raw values and converts them to a scale of 1 to 100.) Originally I had set the loop to "less than 20" and the program did not work. For the life of me I could not figure out why. Then I realized the sound sensor was sensing the noise from the motors as greater than 20. Remember, robots deal in values, not real world experience.

The robot doesn't distinguish between a sound sensor returning a value of 35 from the motors or a student yelling: 35 is 35.

The program begins with a "wait for" icon, waiting for a loud sound, in this case greater than 90. You cannot set this sensor to greater than 100 as 100 is the highest value it will read. Once it hears a loud sound, it starts the loop. It does something else besides just sitting there. In this case, the something it does is drive forward. I turned both drive motors on.

If you turn motors on for a certain duration, and then have them wait to sense something before they stop or turn or do something else, you may just overshoot what you're trying to sense. For example, suppose you are looking for the end of a table so your robot can stop before it falls off. If you program your robot to drive for five rotations, then look for a large drop with your ultrasonic sensor, it may just drive right off the table before it gets to the "wait for distance" command.

In the example provided, by turning the motors on for no particular duration, the robot will drive forward smoothly as long as there is noise greater than 50. If you look at the loop in this program, the sound sensor loop is set to less than 50. This means it will continually loop until the sound drops below 50. If the motor was turned on for one rotation, you would see the robot move forward and stop, move forward and stop, continuously until the noise level dropped below 50.

Can you think of a simple way to start this program over from the beginning if the noise level does drop below 50? See Example Two on page 86 for the answer.

Color Sensor

The classic program with a light sensor is to have a robot drive forward and stop on a black line. Because of the reflective difference, black tape on a white surface works really well for this. I use tape on a white surface for a number of different challenges: stop on a black line, stop on a specific color line, line follow, say the name of a color, and a few other things.

GO TO THE TAPE

Having some tape readily available can be helpful in a classroom or in a build room. If a student is passing every challenge with ease, a new challenge can be created quickly by adding a new color, an extra line, or by bending a straight line to make line following more difficult. Colored electrical tape can create hundreds of challenges, and they can be created on the spot. That can't be done on pre-printed mats.

Let's start by having your robot drive forward and stop on a black line. Let's skip over the algorithm and pseudo-code here for brevity and clarity. If you still find that you need these to help you out, you can try to write them down anytime you wish.

To write a program, two pieces of information are needed: the reflection of the tape and the reflection of the white surface. Once these two numbers are known, it is possible to calculate a threshold, which is a medium number that if crossed, will trigger the next event. The color sensor on the VEX IQ reads reflected light in what is called "grayscale" and provides a value of 0 for total darkness and 400 for total brightness. When the displaySensorValue program described earlier is used, a reading of 15 for the tape and 285 for the white surface is registered. Adding these two numbers and dividing the sum in half produces a threshold of 150.

Here's the program in both RobotC Graphical and Text:

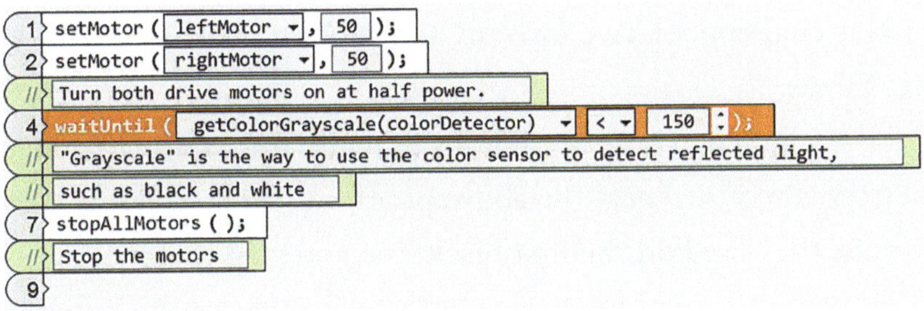


```
task main ()
{
setMotor (leftMotor, 50) ;
setMotor (rightMotor, 50) ;
// Turn both drives on at half power
waitUntil getColorGrayscale (colorDetector) <150) ;
// Grayscale is the way to use the color sensor to detect
    reflected light,
// such as black and white
stopAllMotors () ;
//stop the motors
}
```

RobotC automatically translates graphical code into text, with documentation shown in green.

So if my robot is driving and senses a number lower than 150 it should stop. Why do we use this threshold? Why not just say less than 285, since that would include my black tape, which has a reading of 15? The answer is shadows, bumps, cracks, dirt, etc. There are so many things that can return a lower value than our bright white but are not actually our black tape. By using a threshold between my light and dark values, these false readings are avoided.

Before moving on to using our sensor in color mode, let's try this challenge: Drive forward and stop on the third of three black lines.

This challenge pushes students to really think. The first idea they try is to just repeat the above programs three times. If they do this, they land on the first black line and stop there. They think there is something wrong with their robot or the lines and can't figure it out. This is because they are not thinking like robots. Here's an exercise to put them through to show them where they're going wrong. I have them act like life-sized robots. I have them stand about 5 feet (1.5 m) behind the big yellow rectangle on the floor in my classroom. I tell them to walk forward and stop when they see the yellow rectangle. And they do so. Then I tell them to do it again and they make the minutest of movements until they realize they are already there. Just like their robot is. Their robots don't know the third piece of tape, they only know the third time they see less than a given value (150) and then stop. They do this with blinding speed, making us confused.

So how would you write a program that overcomes this tendency of the robot to get stuck on the first piece of tape with a reflection less than 150? Turn to Example Three on page 87 for a solution.

Let's now switch away from black and white to the world of color. When you're driving on a highway and look out the window at the lines, what color are they? On most American highways, they're yellow or white. And the different colors mean different things. Yellow indicates a separation of traffic going in opposite directions, while white separates lanes of traffic going in the same direction. If you have a broken yellow line on your left, you are allowed to pass, *with caution.* But if you have a solid white line you must stay in your lane and cannot pass.

This creates a problem if you are building a self-driving car because there is only a tiny difference between the reflective light measurements for yellow and white. In a test of a robot driving on a black surface with stripes of white and yellow tape, the EV3 gives a reading of 15 and 18. I would be reluctant to rely on a reflective difference of 3 between the two colors if the safety of my passengers was at stake. One bright, shiny ray of sun and you might be crossing the wrong line at the wrong time. We have another option, however, and that is to use the color sensor in color mode, not reflective mode.

With the color sensor in color mode, we get readings of 0 for the road, 1 for yellow, and 6 for white. These values may seem even closer to you, but in reality they are far apart. Because the values are a measure of color and not reflection, the numbers are not as likely to change. Big caveat: you must check your colors ahead of time. There are only seven preset values EV3 sees, and three or twelve that VEX IQ can read, depending on the mode. I wouldn't let my self-driving car loose on the road without validating the color readings first.

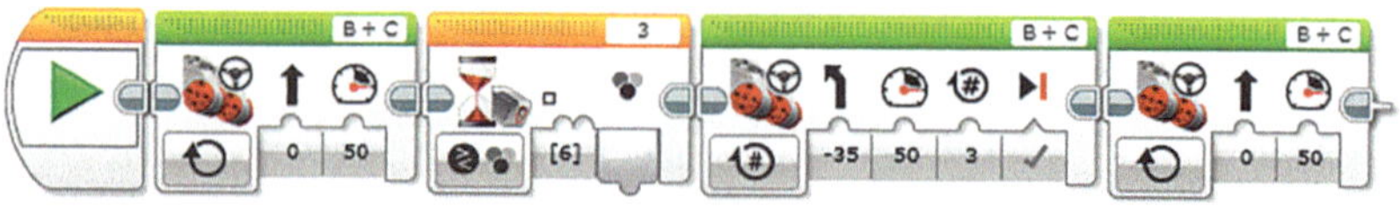

This code uses the color sensor to look for the correct color line so the robot knows it's legal to change lanes.

Remember, "Do something, sense something, do something else." In the case above, we'll drive straight ahead, check the line color, and if it's white we'll change lanes and then resume driving straight.

In chapter five we'll try to figure out how to check to see if the white line is a dashed line or straight line—in other words, whether it's legal or not to cross it.

Basic Line Follow

There are many ways to do line following with a robot. The best one to use depends on the situation. Remember, form follows function. We'll cover a few options here, and in chapter five we'll look at how to smooth out our line follower to make it more efficient by using two color sensors.

RobotC makes line tracking easy as it contains a function that has all the individual commands built in. It's called line tracking, and all you need to do is drag it over into a forever loop. Let's take a look:

```
repeat (forever ) {
    lineTrackLeft ( colorDetector , 160 , 75 , 10 );
}
```

RobotC uses many functions to combine commands. This line tracker command is a function comprised of several individual commands.

There are four variables the programmer needs to fill out here. The first one tells the robot which sensor you are going to use for line tracking. Hopefully you're using a color sensor; if you have more than one on your robot, make sure to choose the name of the correct one—you named them in Motor and Sensor Setup, remember? The next box over asks for the threshold, the average number between the surface your robot is on and the line it is following. After that, it asks for the speed (in percentage) of the motor that will be turning faster, and the final box asks for the speed of the other motor.

The reason you have two speeds is because most basic line following is a zigzag. You are not actually following the line; instead you are turning towards and away from the line repeatedly while you move forward. That is why this function is called "track" not "follow." Adjusting the speeds in these last two boxes will determine how fast you move forward while following the line. If the speeds are close (but never even or you'll drive straight) you will only turn slightly. This is fine if you are following a straight line. But if the line is curved or turns, you will want one motor on while the other motor is off or a lot slower. This allows you to make a pivot turn, which will help you find your line as you slowly move forward.

Notice that the function says "lineTrackLeft." To use this function, you will have to place your robot with the line to your left. The first thing the robot will do is turn to the left until it sees a reflection that is 160 less than it is currently seeing. If it turns the wrong way, it will almost spin 180 degree before it finds the line.

Most likely nobody will follow a line infinitely, so you can change the "repeat forever" function to one where it is looking for another sensor reading. A common way this is done in robotics competitions would be to follow or track a line until you are a certain distance away from an object. For example:

```
repeatUntil ( getDistanceValue(distanceMM) <= 10 ) {
    lineTrackRight ( colorDetector , 160 , 75 , 10 );
}
```

In this example, you will track a line on your right until the ultrasonic sensor returns a reading of less than or equal to 10.

There are many different variations of line following programs. Because we have not yet used a switch in EV3 (known as an if/else statement in other languages), I'll show you a line follower using a switch that will behave almost exactly like the previous one.

For the following example, I used a robot on a white surface with a piece of black tape. The robot read the surface as 24 and the black tape as 6. Therefore I calculated my threshold as 15. The switch block gives you two choices to follow—one if a statement is true, one if it is false. In this example, *if* the color sensor in port 3 senses a reflection less than 15, then execute the top set of commands (denoted by a check mark for true); *else* the color sensor in port 3 does not see a reflection less than 15, then execute the bottom set of commands (denoted by an X for false). The true line of commands will turn my robot to the right because B is my left motor, and the false line of commands will turn my robot to the left because C is my right motor. It's

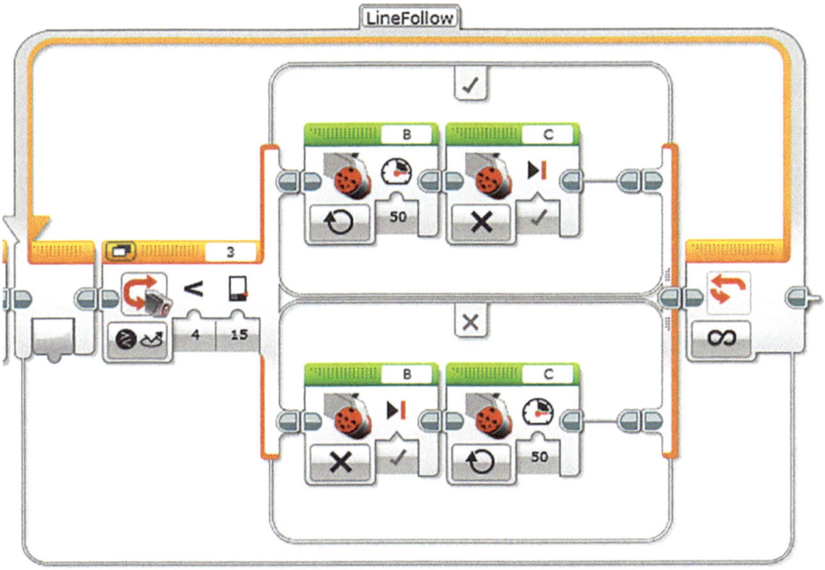

This is a simple line follower for EV3 that uses a switch. The robot will follow the commands either on top or the bottom, depending on conditions.

important to turn one motor on and the other off. I have seen lots of students make the mistake of only turning motors on, and then wondering why their robot is driving straight.

Also be sure to use the Motors On mode and not on for a duration. By just turning them on, they will constantly repeat the loop, checking for the sensor readings. If you turn them on for a particular duration, they might drive over the line before you have a chance to look for it again.

Gyro Sensor

In order to use the gyro sensor correctly, you must understand what it can and cannot actually do. One line in the EV3 Help section can really clarify the problem. It reads: "In addition, the

SENSING DANGER

Gyro sensors can be used to help engineers. Suppose you're building a bridge and want to know how much it sways in the wind and whether its waves are up and down or back and forth. You could mount two gyros to detect each type of motion you are looking for (up/down and tilting left/right). Instead of driving or moving, you would reset each gyro, and then let the robot wait for an hour during a storm (affixed safely, of course). After the hour, you could read the display to see how much change in angle each gyro reported.

This is not a hypothetical situation. The collapse of the Tacoma Narrows Bridge is one of the most famous bridge disasters in history. Galloping Gertie, as it came to be known, had always swayed in the wind, but in 1938 it swayed a little too much and came crashing down. Engineers need to know this sort of information about their designs so they can build safer infrastructure for our country.

If a sensor had been on the Tacoma Narrows Bridge, we would know how much stress it handled before it collapsed.

Gyro Sensor keeps track of the total rotation angle in degrees. You can use this rotation angle to detect, for example, how far your robot has turned." This last part says you can detect how far your robot has turned, but it does not say you can program your robot to turn a certain amount using the gyro sensor. Except sometimes you can, and some tutorials encourage you to do this. You just have to turn slowly, get lucky, and account for drift and other factors.

So let's take a look at a few basic situations. Begin by having the gyro sensor record how much your robot has turned, like the example in the quote above. This program should do the trick:

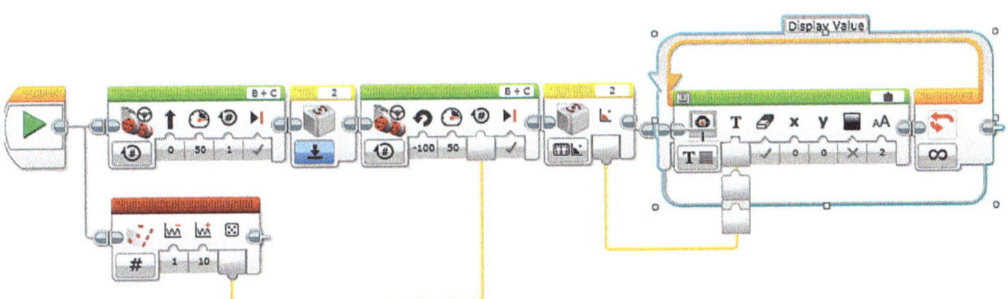

This program records and displays the change in an angle as detected by a gyro sensor.

This looks complex, but we can break it down into easy-to-understand parts. When the program begins, we are driving forward for a single rotation and also generating a random number between 1 and 10 (red icon). Then we reset the gyro. Since I am not waiting for a specific event, I have taken the gyro sensor block from the yellow palette, not from the orange one. After resetting the gyro my robot will spin for a random number of times as wired in from the random block. When it stops, I read the change in angle and then wire that value into a display block in a forever loop so it doesn't disappear on my screen.

So why is this important? It seems like a pointless program to record the angle of a random turn. Well, what if it wasn't random? What if we continually changed the duration of the turn until the program returned a value of 90 degrees each time? That would be one surefire way to know how to program our robot to make a 90-degree turn. If we know this ahead of time, we can calculate any angle turn on a competition field just by using 90 degrees as our benchmark. Other methods are not as reliable.

You can program the motors to turn really slowly until the gyro sensor detects a change in angle of 90 degrees. However, this method is iffy because, as the EV3 help guide says, it is accurate within +/– 3 degrees. Too many factors, including speed of turn and drift, can affect your values. A similar program in RobotC faces similar difficulties.

Summary

We have just gone over several different ways to use sensors. There are many more ways to use them to accomplish simple tasks and even more advanced uses. In each programming language, you could write several programs that would accomplish the same task. The possibilities are nearly endless, but the fundamentals are not. Here they are: You must use the right sensor for the job and attach it in a manner most efficient for giving you the values you are looking for; learn how to read your sensors using Port View or writing a simple program to display values; take your time so you can see things how a robot "sees" them; and finally, "Do something. Sense something. Do something else." This works for the simplest of programs but can also be applied to more complex tasks with a little bit of modification.

Problems Solved

In this chapter there were three problems presented for which no answer was provided. Programming and building is a creative endeavor in which there are many solutions to a given problem. Here is one solution for each of them. If your solution doesn't match this solution, that doesn't mean you're wrong. It means there are different solutions to the same problem. And there's nothing wrong with that.

Example One

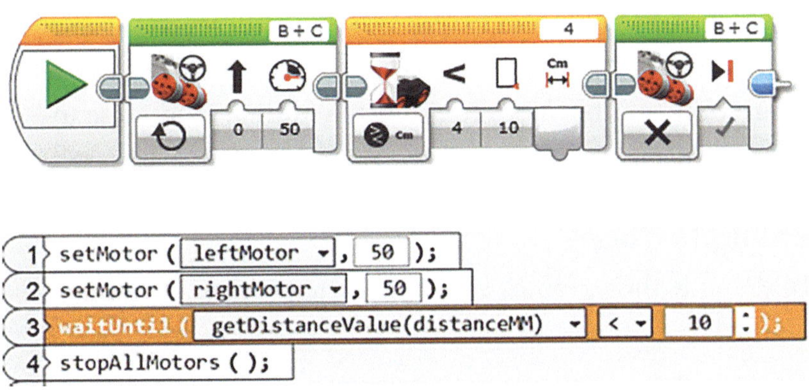

Here is how to use the ultrasonic sensor to stop 10 centimeters (4 inches) in front of a wall in two different graphic languages.

The first question asked how you would program your robot to drive forward and stop, using a distance sensor, 10 centimeters (4 inches) from a wall. On the previous page is a solution for both the EV3 and VEX systems.

Example Two

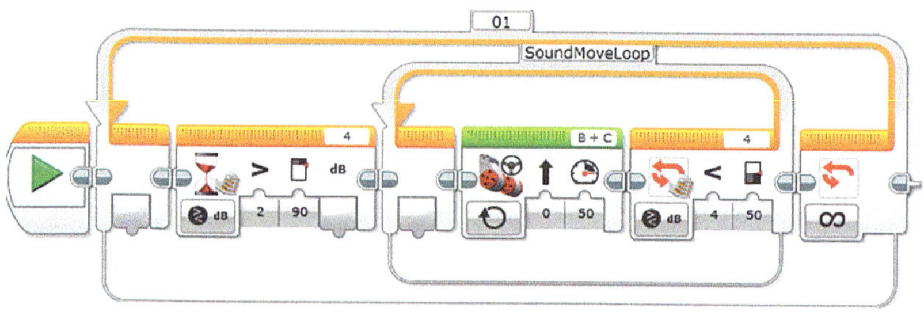

By placing the original program in a loop you can start over by making noise once again after the robot has stopped

In example two, there was a problem in which the robot was programmed to move continuously as long as a loud noise was heard. The question was then asked if you could think of a simple way to start this program over from the beginning if the noise level dropped below 50. One answer is to place the entire program in another loop. It will listen for that loud noise to start again:

Example Three

In example three, robots were being programmed to stop on the third of three black lines. The problem was to write a program that overcomes the tendency of the robot to get stuck on the first piece of tape with a reflection less than 150. Here is one solution:

```
Task main ()
{
        repeat (2) {
                setMultipleMotors (50, rightMotor, leftMotor) ;
                waitUntil (sensorValue [light] < 150) ;
                waitUntil (sensorValue [light] > 150) ;
                }
                setMultipleMotors (50, rightMotor, leftMotor) ;
                waitUntil (sensorValue [light] < 150) ;
                stopMultipleMotors (rightMotor, leftMotor) ;
```

By having my robot drive forward and look for dark, then light and repeat this two times, I get it past the second piece of black tape. Then it just drives forward one more time, looking for black, and stops.

All programming blocks are downloaded into the robot's brain from a computer. Choosing correctly can require teamwork.

5 Where Do We Go from Here?

One of the most exciting features of the EV3 is the ability to use third-party sensors. The first thing you will need to do to actually use them is to download programming blocks to use with your EV3 programming software. Different companies have approached this differently, and I'll cover two of them here to show you some of that variability.

HiTechnic made sensors for the NXT and now makes them for the EV3. To cover the EV3, they have simplified their sensor blocks so they have one for their analog sensors (those that return an infinite range of values) and one for their digital sensors (those that return values in discreet increments).

Navigate to the HiTechnic download page (www.hitechnic.com/downloads) and choose the link for EV3 programming blocks. Download the one that you want. It will download as a .zip file. From that file you must extract the folder by double-clicking on the .zip file. Then open the folder to show the different sensors available.

Once you have identified the sensor block you want, open up EV3 and start a program. The option to import a block will not work if you try to do it from the lobby. Once you have started a new program, click Tools>Block Import. Then navigate to the block you want and choose it. In the example below, I chose the compass sensor. You will have to restart the EV3 software in order for the changes to take effect.

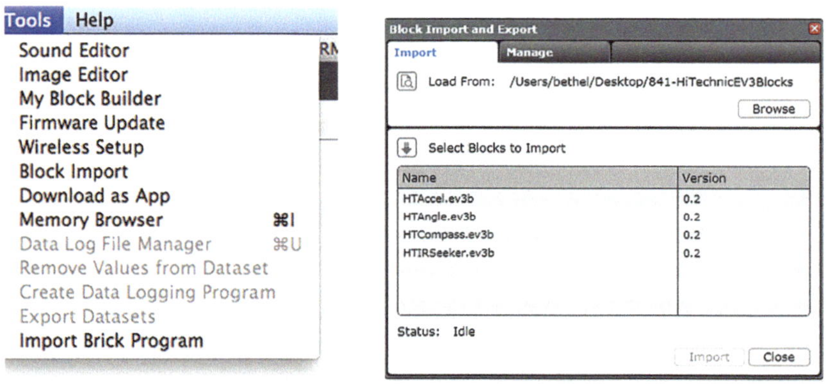

Choosing the Block Import Tool under the tools menu will allow you to import the correct programming block.

Once you have restarted the software, you won't find the new block as an option in the "Wait For" blocks found in the orange "Flow" palette. This is because that would be modifying an existing LEGO Education EV3 block, and they don't want you doing that. Instead you will have to look in the yellow "Sensor" palette to find your new block.

I have already shown you how to wire information from a sensor when we output gyro readings; but what if we want to use it in a simple program like we would with any other "Wait For" block? One popular option is to create a program that does what we want and then to wrap that program into a My Block. My Block is a great function of EV3 that allows you to take code you use often and wrap it into one individual programming icon. It saves memory in the robot and space on the programming palette. Let's take a look at how it's done. First we'll create a program that waits for the sensor to read a heading of 360 degrees, or north. (This is not a measurement

True north generates a reading of 360 degrees in a compass sensor. This car needs to make a big turn to head north.

of angle like the gyro sensor, but a true magnetic reading of the compass directions.)

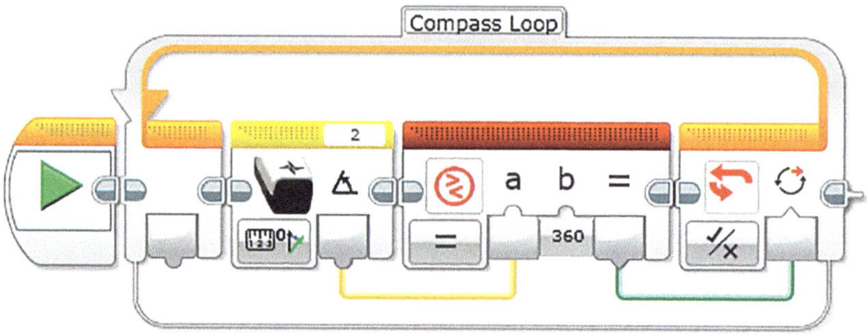

Here's a way to tell our robot to wait for its compass sensor to sense north.

This is a program with a logic loop. The loop will continue repeating until it receives a "true" signal. The true signal will be output from the red compare block. This block will only output a true signal if its conditions are met. I've set those conditions so that *a* will equal *b*, which I've set at 360. The reading for *a* was output from the yellow compass sensor block, which I set to the "Measure—Absolute Heading" setting. If you put it all together, the compass sensor continually checks its heading and outputs that to the compare block. The compare block outputs a false signal until the reading it receives is equal to 360. When it is equal to 360, it outputs a true signal to the loop, which then ends. In other words, wait for 360.

This is useful and will work, but I would prefer to have one block that does this. And I would like to be able to change the heading at which I want to stop. Both of these are achievable with a My Block. First, highlight the part of your program you wish to make into a My Block. In our example, it would be everything except the start icon—you can never include these in My Blocks. Then, from the Tools option at the top, choose My Block Builder. You will get a pop-up window that looks like this:

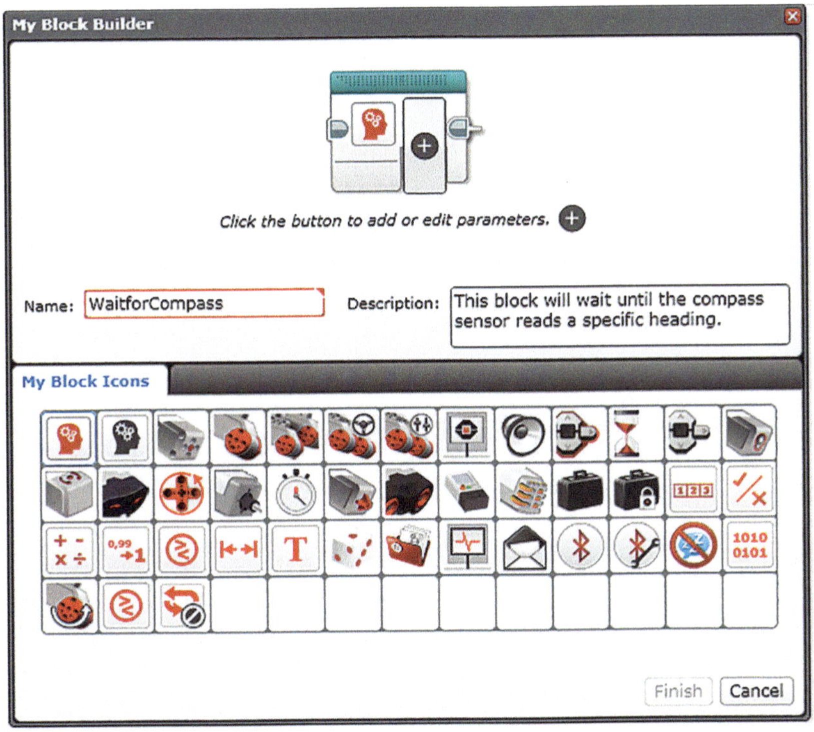

When building a My Block, you will be asked to name and describe your block, and to choose an icon for it.

The My Block Builder asks that you give your block a name and a description and asks you to choose an icon for your block. I've done that and left the icon to the default red man with gears in his head. That's the basics of my new block, but I also want it to work for choosing any heading between 1 degree and 360 (the compass sensor doesn't like 0; it prefers 360 for a heading of north). To do this, I need to click on the plus sign at the top where it asks me to add or edit parameters.

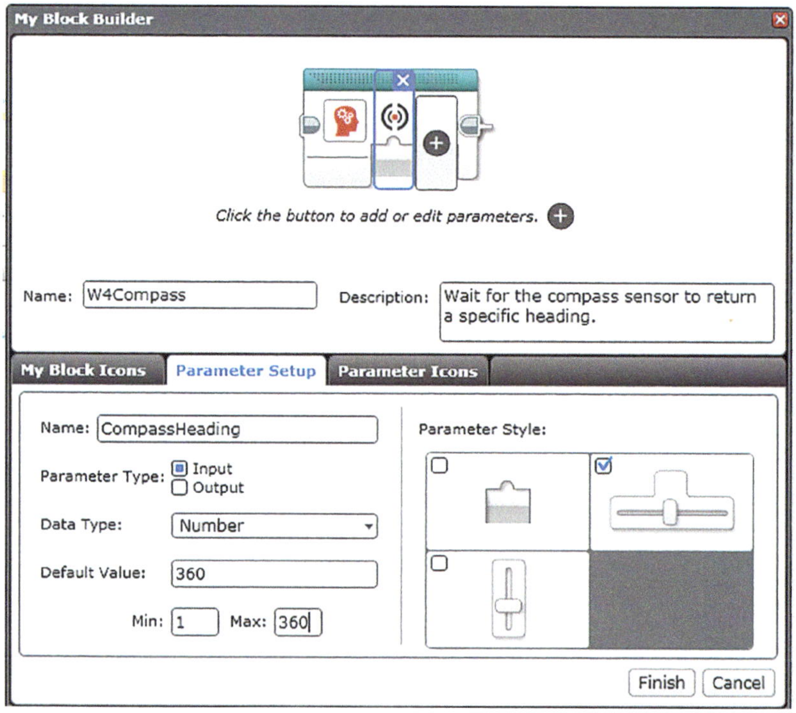

You will have to make many decisions when editing paramenters for your My Block.

If you've made it this far, you can choose your own icon for the parameter. You can give your parameter a heading, choose if it's an input or output, and then choose the type (number, text, logic, etc.). Finally, you can choose a slider or text input and provide maximum and minimum input values, plus a default value. After all that, click "Finish" and you're not quite done. My Block Builder needs to know what value in the original program the parameter is going to change. So the parameter will appear next to the original program, and you have to drag a wire from it to the value you want to change, in this case it's the value in the compare block (*b*) that I am comparing my compass heading to.

And now I've got a new block ready for my use with any program that needs me to wait for the compass sensor to return a specific heading. There is one final consideration when creating a new programming block with the My Block Builder. Your new block is available to use in the light blue "My Blocks" palette. However, it is only available to any program in the current project. In order to use it in another project, you need to export it to a folder (where hopefully you'll save lots of your own created blocks) on your computer somewhere. To do this, click the little wrench next to the name of your programs in EV3. This will bring you to Project Properties. From here, you click the tab for My Blocks and then export the appropriate block.

Creating a My Block takes some time to get right, but it is an invaluable tool for any programmer who wants to go beyond the basics. It will ultimately save you a lot of time and robot memory. As you build your programming skills, you will really want to use this tool more and more often as it also helps you really understand what different functions are doing.

My Blocks are specific to EV3; in most other programming languages they are referred to as functions. Each program works differently, but the basics are the same, you name the function, define the function, then call on the function when you are ready to use it. In programming, functions are popular as they save time and space.

If you want to get into scientific experiments with sensors and probes, then Vernier (www.vernier.com) is the place to go. To use their sensors, the programming block is downloaded off their website and imported exactly the same way as the HiTechnic blocks are. However, there are two important differences. The Vernier sensor block is a single block that can be configured to use with any of their twenty plus EV3-compatible sensors.

There are many companies that make third-party sensors for the EV3. It's important for you to look around and find the sensor you want, experiment on your own, and see what you can do.

Timing a Drag Racer

There are two basic types of sensors your robot has that we haven't discussed. You don't have to attach these sensors to your robot; they are already there. The first one is the timer, which we use anytime we want to record the duration of an event or want to use it in its "wait for" mode to have an event continue for a certain amount of time. The second one is the motor encoder, which is a little electromagnetic device contained inside the motor. It can read the motor speed and amount of rotation. This information will be useful in the next programming challenge we are going to inspect.

Here's the problem: My students need to have an accurate time for their drag racers. Racing them against each other makes it hard to tell who won when it's close and pushing a button on a smartphone timer isn't very accurate. The exact time is important because in my class an A is under 3.5 seconds and a B is 3.501 to 4 seconds.

The solution we came up with is to use the natural brightness of the big yellow rectangle I use for the drag racers. By placing black tape at the end of the strip, we have a large difference between the amount of reflective light at the start and finish.

Because each student team used different sized wheels, we couldn't use the encoders in the motors to measure rotation, so we had to use reflective light to determine when the dragsters had crossed the finish line. Here's the program we worked out:

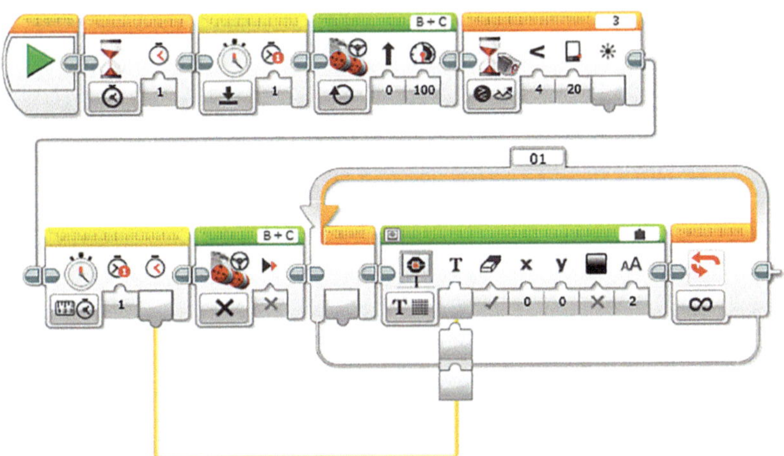

In this program for timing a drag racer, the robots wait for one second before racing at full speed.

In the program above, we wait for a second before we start. This gives the students time to take their fingers out of the way after they touch the start button. Then the timer resets and the drag racers take off at full speed. When they sense the black line (reflection less than 20), two things happen: One is the timer is read and the output is wired into a display block. The second is the motors are turned off. Notice that most buttons on the Motor block have an X. The first X indicates that the motors are turned off. The second one indicates they are set to coast, not brake. If they were to brake, they would stop exactly at that moment and lock up, which can break the gears. By coasting, power is shut off to the motors and they are allowed to slow down as momentum dictates.

Using Two or More Sensors

Both EV3 and RobotC Graphical are powerful languages, but there comes a point where you run out of room for graphics and you need text. In addition, text files take up less memory than graphical files do. We'll use some RobotC text-based programs to answer a series of challenges related to a self-driving car

```
taskmain ()
{
        setMultipleMotors (50, leftMotor, rightMotor) ;
        if (getDistanceValue (distanceFront) < 50) {
                if (getColorName (colorDetector) == colorRed) {
                        stopAllMotors () ;
                } else {
                setMultipleMotors (50, leftMotor, rightMotor) ;
                }
        } else {

        setMultipleMotors (50, leftMotor, rightMotor) ;
        }
}
```

This is how to program a vehicle to stop at a red light using a RobotC text-based program. The vehicle will keep going if the light's not red.

approaching an intersection. Let's start with the first step: stop when you sense a traffic light.

For this challenge we'll assume the robot has an ultrasonic sensor mounted on the front facing upward, looking for a traffic light, as well as two mounted on the sides, looking for other cars. Finally, there's a color sensor mounted next to the front ultrasonic sensor. At the bottom of the previous page is the program I'd use to stop at a red light.

I wrote two if/else statements to complete this program. The first one checks to see if there's a traffic light with the ultrasonic sensor. If there's nothing there (*>50*), then the motors keep going. If there is something there (*<50*), then the color sensor checks to see what color it is. If it's red, you stop all motors. If not, then you keep going.

For the next challenge, let's add a new condition for the traffic light, red or yellow. In this case I'll just add another if/else statement to check on the color of the light if it's not red.

By adding a second if/else statement (see page 100), we can account for the light being yellow if it's not red. In this case, I have the motors slow down when the light is yellow, otherwise they continue on at the same speed. (The *noMotorOnPort* appears because the default for *setMultipleMotors* is to set the value for four motors.)

Finally, I've added one more level to the challenge (see page 101). If it is a green light—not yellow or red—you have to account for traffic that might be coming from either side. Remember the goal of the self-driving car is to be safer than a human driver. Otherwise consumers won't go near them. So even when there's a green light, the car is going to check

```
taskmain ()
{
  setMultipleMotors (50, leftMotor, rightMotor,
  noMotorOnPort, noMotorOnPort) ;
  if (getDistanceValue (distanceFront) < 50) {
        if (getColorName (colorDetector) == colorRed) {
              stopAllMotors () ;
        } else {
  if (getColorName (colorDetector) == colorYellow) {
        setMultipleMotors (20, leftMotor, rightMotor,
  noMotorOnPort, noMotorOnPort) ;
        } else {
        setMultipleMotors (50, leftMotor, rightMotor,
  noMotorOnPort, noMotorOnPort) ;
        }
  }
} else {
  setMultipleMotors (50, leftMotor, rightMotor,
  noMotorOnPort, noMotorOnPort) ;
  }
}
```

Using these instructions, the car will top at a red light, slow down through yellow, and go through green.

to make sure no traffic is coming from the left or right. In the example to the right, I have shown a program checking only one side, but you can probably figure out by now that you just need to add another if/else statement to account for another condition.

Can you spot one mistake made in the line for the distanceLeft sensor in the example on page 101? If you think you got it, please check Example Four on page 113.

```
taskmain ()
{
        setMultipleMotors (50, leftMotor, rightMotor,
          noMotorOnPort, noMotorOnPort) ;
        if (getDistanceValue (distanceFront) < 50) {
                if (getColorName (colorDetector) == colorRed)
{
                        stopAllMotors () ;
                } else {
                if (getColorName (colorDetector) ==
                  colorYellow) {
                        setMultipleMotors (20, leftMotor,
                          rightMotor, noMotorOnPort,
                          noMotorOnPort) ;
                } else {
                 if (getDistanceValue (distanceLeft) == 20)  {
                 stopAllMotors () ;
                } else {
                setMultipleMotors (50, leftMotor, rightMotor,
                    noMotorOnPort, noMotorOnPort) ;
                }
        }
        }  else {
        setMultipleMotors (50, leftMotor, rightMotor,
          noMotorOnPort, noMotorOnPort) ;
        }
}
```

Using these instructions, in addition to responding to the traffic light, the car will check for oncoming traffic approaching from the left side.

Here's another real-life example involving our self-driving car for the next challenge. The object is to make a car recognize whether a white line is broken or solid. That's important because it is illegal to cross over a solid white line, but you can cross a dashed one. You can figure out from the program code in the previous challenge how to check to see if there's traffic, and we already know how to distinguish between a white and yellow line, so we're just going to concentrate on determining if the white line is dashed or solid. This means our program will really be a fragment of a program and not the entire thing. I'll give you the algorithm and pseudo-code, then see if you can write the program and check Example Five on page 114 to compare your solution to mine.

Dashed Line Algorithm

1. While driving:
2. See a white line.
3. Check length of white line.
4. Compare how long white line lasts with known time for a dash.
5. If white line is seen for too long, go back and wait for next white line and check again.
6. If the white line is not too long, turn on signal, change lanes, and keep driving.

Pseudo-code

1. Drive forward continuously.
2. Wait for color equals white.
3. When color equals white, compare duration of white to 1 second.

4. If color equals white for more than 1 second, continue driving, repeat program from beginning.
5. Else turn on turn signal for 3 seconds.
6. Change lanes.
7. Continue with next part of program.

There are many factors to consider that are not a part of this program. Do you even want to turn? Is there anything else in your way? Is there a stop light ahead that is red, meaning you should slow down instead of changing lanes? There are so many variables to consider that it becomes a large task. That's why there are teams of engineers working at places like Google and NASA and each member or group writes little bits of code. Nobody writes the entire program for a major undertaking anymore.

TRACKING TIPS

When tracking, make sure your color sensors are shielded properly, and make sure they are far enough apart that they straddle the line you are following.

So you think you're up to writing code for the pseudo-code above? Try it, then compare it with Example Five to see if your solution looks like mine does. Keep in mind, we may not match and you still may have come up with an acceptable, or even better, solution!

Using Two Color Sensors to Follow a Line

Remember that line following is really tracking the edge of a line and that the robot usually did so in a zigzag fashion. This is true when you are using one color sensor. If you were to attempt the challenge of line following with two color sensors, you will find an elegant, logical solution that has your robot driving smooth as can be. While this looks cool, it also has practical applications. For example, in a robotics competition, if you are following a line and are trying to use a sensor to help you stop somewhere along the line, it helps if your robot is not shaking back and forth as it tracks the edge. Using two color sensors can help you do this.

In this example, we are going to assume the two colors on the side of the line are the same. You can write a program accounting for different colors or reflections, but we'll start with the simple case.

Assume we're following a black line with a reflective value of 10, with a bright white surface on either side with a reflective value of 70. This gives us a threshold or midpoint number of 40. We'll begin with naming our color sensors, left and right. There are four conditions we can look for with these two sensors. Can you guess what they are? Left sensor sees white (>40) and right sensor sees white (>40). In this case, both sensors see white so you want to drive straight forward and we can keep all four motor speeds constant. However, this is hardly ever the case, as we all know our robots do strange things. So let's examine the other conditions. Left sensor sees white, right sensor sees black. If this is the case, my robot is wandering too far to the left and I want to correct it back to the right. If the opposite case is true, the left sensor sees black while the right sees white, then I have to correct in the opposite manner. The last case is a strange one. If both sensors see black it means you are driving perpendicular to the line you are trying to follow. This is not good and it is hard to correct for because you don't know which side of the line you are on or which way you have to go to correct your mistake.

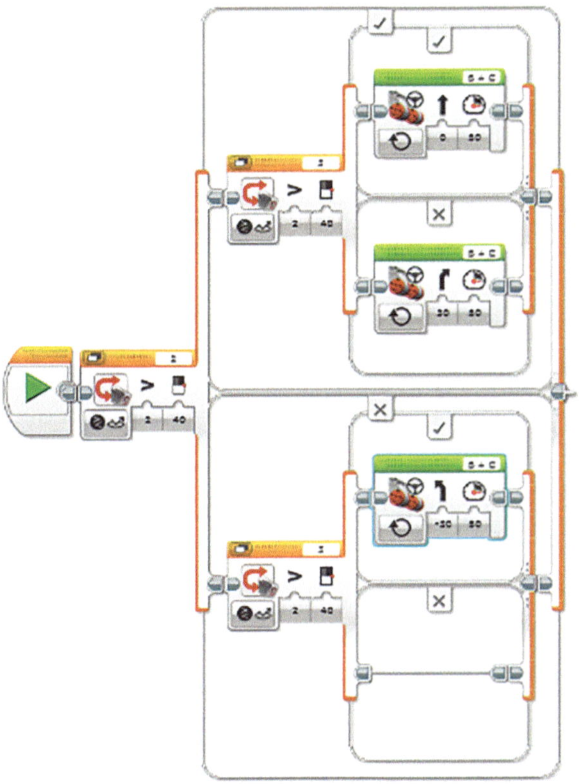

This program is for a simple switch for a robot with two color sensors to follow a line. There are instructions for three of the four possible scenarios.

Here's a simple breakdown of the situation:

L sees White,	R sees White	= Drive straight.
L sees White,	R sees Black	= Turn right.
L sees Black,	R sees White	= Turn left.
L sees Black,	R sees Black	= You're in trouble.

The program for this is pretty simple and straightforward:

In the image on page 105, the color sensor in port 2 is our left color sensor and the one in port 3 is our right one. I have used two switches within a switch to account for the four conditions we outlined above. They read from top to bottom the same way my breakdown just after the program did. I left the bottom space blank to account for not really having an acceptable solution if both sensors see black. Based on the individual challenge you are working on, you may know what to do if your robot is facing the line, but without more information, I can't being to guess how to get it back on track in this generic example.

There's one final improvement we can do with our line follower, and it will make you the envy of all your friends, fellow students, and other robotics teams.

Proportional Line Follower

You will have to play with the numbers being provided to make this work for your robot and your line, but the general concept is accurate. A proportional line follower will only have you correct your turn in proportion to how much that turn

have a slight right turn. This should supply the correction for the left veer of the robot.

Scenario Two: Our robot is driving straight. Both sensors send a value of 70. The difference between the two is 0, and 0 multiplied by our gain factor of 0.5 is 0. No correction for steering will keep us driving straight.

Scenario Three: Our robot has turned sharply to the right. The left sensor is reading 10, while the right is reading 70. Making sure we subtract the right from the left, our difference is –60. Multiplied by our gain, this gives us a steering correction of –30. This is a bit stronger than the small correction in scenario one, and since it is negative, it will turn our robot to the left.

What we have just created here is an example of a proportional-integral-derivative, or PID. PID is an industry standard corrective feedback loop for running all sorts of machinery. A proportional–integral–derivative controller (PID controller) continuously calculates an error value as the difference between a measured process variable and a desired set point. The controller attempts to minimize the error over time by adjustment of a control variable.

In this definition, the error value is the difference between our two light sensors. The control variable is our error multiplied by the gain. Each time we run our loop, the errors should get less and less, thus providing the corrective feedback—that is, until our line changes directions.

Program a Straight-Driving Robot

Anyone who has taught or coached robotics for any length of time will have learned that the darn things just don't go straight. There may be some slippage in a wheel, motors may not be in sync with each other, the robot may be weighted to one side, the build may be asymmetrical, a tire may be worn, the surface may slant, and a thousand other factors can cause the robot to veer. This is especially frustrating in competitions when there is no line to follow or line up against and you want your robot to drive straight to a challenge module.

The gyro sensor was not covered extensively in chapter four because it takes a lot more effort and understanding to create a simple gyro program. With three blocks, I can program my robot to drive forward and stop on a black line. I can use three blocks to make it drive toward a wall and stop 20 centimeters (8 inches) away. I can also use three blocks to make it drive and stop when the touch sensor is pushed in. But three blocks will not be enough to program my robot to turn 90 degrees and stop using the gyro sensor. That is the reason why it is overlooked for most beginner applications. Now that beginner applications have been covered, we'll have a go at programming the gyro sensor.

First, we need to understand the two factors that cause a lot of fuss in the first place: drift and **lag**. I have mentioned drift before; it is the tendency of the gyro readings to start increasing even when the robot is still. What is actually happening is the gyro is trying to calibrate itself. It wants to orient itself to its environment when you turn it on. It is expecting to set itself to 0 degrees of rotation in a perfectly still environment, but sometimes we turn it on and it is not perfectly still. So it senses movement as 0, and then, when it is still, it tries to account for the fact that it is not at 0 and the numbers move. You should unplug and re-plug the gyro until this stops. That works but is frustrating. So here is a fancier way of calibrating it in a program:

Set it to measure rate and then measure angle, then wait a second. By switching between these modes while it is still (waiting a second) it will recalibrate itself—no messing with plugs. Just make sure that if you switch in this order, you use the angle mode for the rest of your program. If you switch back to measuring rate at any point, it will recalibrate itself when you don't want it to.

The other factor that affects gyro sensors is lag. This is the tendency of the sensor not to keep up with the movement of the robot. To combat this, try making the robot turn slowly. I have also read suggestions to take a few degrees off of your desired angle. With some experimentation and trial and error, you can probably get very accurate with this.

Now that we've looked at some ways to account for lag and drift, let's see if we can write a program to make our robot drive straight. The secret here is we are going to use a form of PID again. We'll measure our error, multiply by a corrective factor, then apply that correction, and check the system again.

For our final challenge, I need to drive in a straight line for twenty rotations. First, I'll calibrate my gyro sensor using the same method just described. Then, we will get our reading from our gyro. In this case, we don't have to calculate the difference between two sensors. We are trying to drive straight forward, which would be 0 degrees since we just calibrated our gyro sensor. Any variation from that heading is our error. So if the gyro sensor reads 0, we are going straight, and if it reads something else, that is our error. Again, we'll apply our error to a corrective motor turn and then repeat the loop until we have completed our twenty rotations.

The two variables that you will have to adjust for yourself are the speed and the amount of correction (gain) that I applied

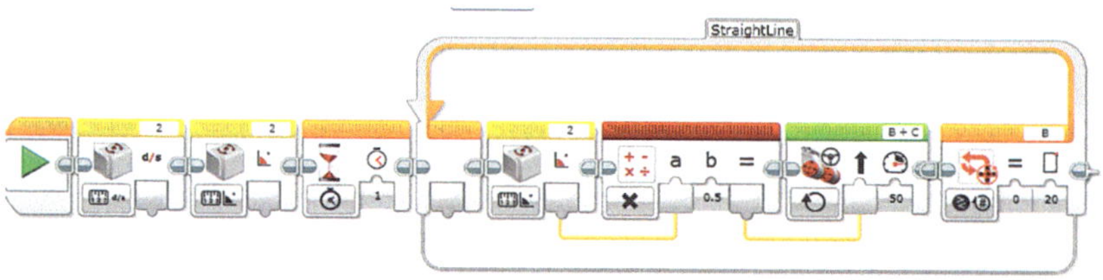

This program for driving in a straight line using a gyro sensor uses a form of PID and starts by calibrating the sensor.

above. I use half power (50) and a correction of 0.5 again because I have found that that works for me. You'll have to decide what works for you. This program will account for drift by calibrating the sensor and account for lag by repeating the loop over and over faster at super high speeds so it is constantly correcting and inching toward zero error like any good program and programmer should.

Summary

Like every other chapter, this one is not meant to cover all the ground out there when it comes to the advanced use of sensors in educational robotics. I've barely scratched the surface. Hopefully I've given you a good start to help increase your knowledge and whet your appetite for more exploration. If self-driving cars (and self-flying planes!) are the future, then the future is going to need a lot of engineers to help build and program them!

Problems Solved

In this chapter there were two problems presented for which no answer was provided. In the first (Example Four), readers were asked to find an error in a line of coding. In the second (Example Five), we are determining if a white line is dashed or solid so our self-driving car can change lanes. An algorithm was provided.

Example Four

The line that reads:

```
if (getDistanceValue (distanceLeft) == 20) {
```

should have a less than (<) sign instead of the equals. Otherwise we'll only ever see the oncoming traffic at that exact distance. If it's already closer than 20 when we start sensing, we'll miss it if we're only looking for distanceLeft = 20.

Example Five

This solution may not be the most elegant, but it will work. Follow the description from left to right: We enter a loop and drive along until we sense white. We then reset the timer and wait for not white. When we see not white, then we record the

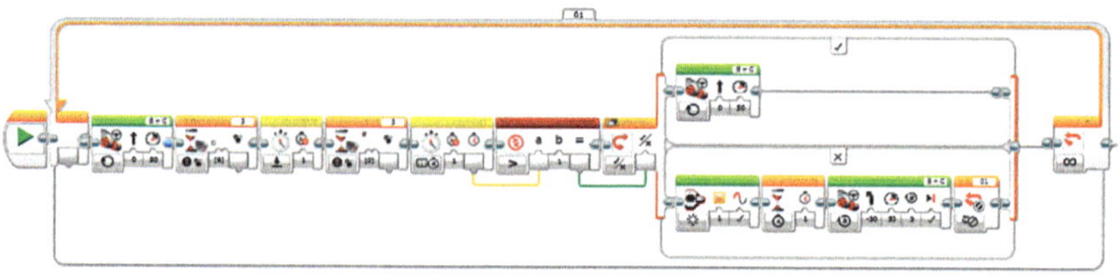

By reading the elapsed time for how long the color sensor saw white, the robot can determine if it was a dashed line or a longer one and then decide to keep going or change lanes.

length of time that passed while seeing white. We compare that number to 1 second. If it's longer than that, we produce a true logic output and follow the top row of the switch, which has us continuing to drive forward and then follow the rest of the loop back to looking for white again. If the duration of white was not more than 1 second, a false logic signal is sent which has us follow the bottom of the switch. This tells us to turn our blinker on, change lanes, then exit the loop using a "loop interrupt" command. At the end of the loop, you could continue with whatever code you wish.

Glossary

acceleration sensor A sensor that can measure tilt in three directions, along the *x*, *y*, and *z* axes.

ambient light The total amount of light in a given area not produced by a specific source.

analog sensor Any sensor that has an infinite range of values it can detect and convey to the robot.

anemometer A sensor that measures wind speed.

angular velocity Velocity is a vector measure, which means it takes into account speed and direction; thus angular velocity measures the direction and speed around a specific axis. Angular velocity measures the speed at which a robot turns around a given axis (*x*, *y*, or *z*) and which way it is turning, so it will also include a positive or negative component.

array An arrangement of items in a particular order, usually spread out; or a container that holds a fixed number of values such as those read from a sensor.

axis A fixed line about which a body will rotate. For example, the Earth spins about its axis. In physical space, there are an infinite number of lines that could be an axis, but for the purpose of robotics sensors, we usually refer to three different axes—*x*, *y*, and *z*—where *x* stretches to the left and right, *y* stretches to the front and behind, and *z* stretches straight up and down. A body can rotate around one or more of these axes at a time.

axle A LEGO piece of different lengths which has a cross profile when you look down one end. Axles are used as connecting pieces to attach gears and wheels to motors or other pieces.

barometer An instrument that detects a change in pressure in the atmosphere.

coefficient of friction A measurement of the ratio between the friction that holds two items together and the force needed to move one of those items.

current probe A sensor that can measure DC currents.

decibel A unit that measures the intensity of a sound wave—basically how loud or quiet a sound is.

digital sensor Any sensor that has values that can be reported in discreet increments.

drift The tendency of gyro sensors (this seems to be true for EV3 and VEX IQ) to return an increasing series of numbers even when the robot and sensor are absolutely still.

echolocation When an object (usually an animal like a bat or a dolphin) determines its location by measuring the time it takes for an echo of a sound (usually a high-frequency wave) to return to the object.

encoder A device inside a motor that determines the speed or position or both of the motor.

engineer A person who is educated and trained in the design and control of engines, machines, and systems; as in a mechanical engineer or software engineer.

gate Part of a sensor that allows an amount of voltage through depending on the reading of the sensor.

GPS Global Positioning System; an electronic system using satellites to determine positions of people, vehicles, phones, etc.

heading The direction in which you are going or from which a signal or object is coming.

icon In programming, a visual block that usually conveys a specific command or set of commands to the robot.

inertial measurement unit A device that can be thought of as a "mega-sensor" that combines gyroscopes, accelerometers, and sometimes magnetic sensors to determine information about an object's force, angular momentum, and linear momentum.

infrared Referring to waves on the electromagnetic spectrum that are invisible to the human eye, shorter than a radio wave, but longer than a light wave.

infrared beacon A device that emits an infrared signal, often in different frequencies.

infrared sensor A sensor that can detect and read infrared signals.

lag The tendency of the gyro sensor to not keep up with the speed of a fast turning robot.

laser The origin of this term comes from shortening "light amplification by stimulated emission of radiation." A laser is a focused beam of light caused by exciting atoms to higher energy phases; the beam produced by a laser does not diffuse as quickly as those produced by other sources.

limit switch A type of touch sensor produced for VEX EDR which has a thin aluminum arm for delicate touch situations.

line follow Refers to the fun educational activity of programming a robot to follow a line; to line follow is an indispensable tool for robotics competitions.

line tracker A sensor produced by VEX EDR for following a line using an infrared light sensor.

orientation The direction in which your robot or sensors are placed in reference to a fixed object. You can't be oriented "straight"; you can be oriented "parallel," in reference to the ground or an axis of rotation.

pH sensor A sensor designed to measure the relative acidity or basicity of a liquid using the pH scale.

programming The act of giving commands to your robot, usually through a programming language like EV3 or RobotC.

proportional-integral-derivative A controlled-loop feedback mechanism that accounts for error, applies a correction, minimizes the error, then checks again and repeats the process.

proximity sensor A sensor that detects the presence of an object; unlike an ultrasonic (distance) sensor, a proximity sensor does not necessarily give you a measurement of how close the object is.

radar A system of detecting and locating an object using radio waves in a manner very similar to how echolocation uses high-frequency sound waves.

reflective light The amount of light that bounces off an object when a beam of light is pointed at that object.

robot A programmable machine built to carry out a task or series of tasks automatically; the term "robot" comes from a 1921 play written by Czech playwright Karel Čapek and referred to humanoid-like machines.

RobotC Programming languages created by Carnegie Mellon University that work for a variety of robots like EV3, VEX IQ, and EDR, and contain graphical and text versions.

rotational motion Movement around an axis; circular movement.

salinity sensor A sensor designed to measure the amount of salt in a liquid.

sensor Any device designed to measure and relay information about observable phenomena.

soil moisture sensor A sensor designed to detect the amount of moisture in soil; it is very useful for gardening.

switch The name of a command in EV3 similar to an if/else statement; a switch compares a value to a default state and chooses one of two courses of actions depending on whether the value is equal to the default or not.

temperature sensor A sensor designed to detect the temperature of a specific object or the temperature in a given area.

third-party sensor Any sensor made for one company's robot that is produced by another company.

turbidity sensor A sensor that detects the cloudiness of a liquid.

ultrasonic (distance) sensor A sensor designed to detect the distance between the sensor and another object.

UVA sensor A sensor designed to measure ultraviolet light.

viscosity Refers to the ability of fluids to resist flow or other forces, commonly seen as the thickness of the liquid.

voltage A measure of the difference in electrical potential between two points.

welding robot A robot designed, built, and programmed to perform welding tasks.

Further Information

Competition Websites

FIRST Robotic Competition

www.firstinspires.org

This site is the clearinghouse for all FIRST competitions from Jr. FIRST LEGO League (FLL) to First Robotic Competition (FRC). You can find rules, registration, and area event information here.

High Tech Kids

www.hightechkids.org

This site, run by volunteers in the state of Minnesota, provides free resources and training for teams competing in FLL and other competitions. It is the first site I went to over a decade ago when I started coaching FLL.

Robotics Education and Competition Foundation

www.roboticseducation.org

The Robotics Education Foundation provides two separate competitions for VEX IQ and VEX EDR robots.

World Robot Olympiad

www.wroboto.org

The World Robot Olympiad is a yearly competition that has dance, football (soccer), and search and rescue categories.

Instructional Websites

Carnegie Mellon Robotics Academy

education.rec.ri.cmu.edu

The Carnegie Mellon Robotics Academy home page provides links to resources for students and teachers for both VEX and LEGO formats.

Damien Kee's Technology in Education Page

www.damienkee.com

Damien Kee has been a leader in educational robotics for over a decade now. He has written several books on robotics and his design for Riley Rover is used in hundreds of classrooms (including mine) around the world. I highly suggest you join his robotics email group (sign up from his site).

EV3 Lessons

www.ev3lessons.com

This amazing site is even more amazing because an FLL team hosts it and its main contributors are teenagers. If you are an FLL team member or a young roboticist looking to enhance your skills or knowledge, this should be one of the first sites you visit.

Ian Chow-Miller's YouTube Page

www.youtube.com/ianchowmiller

There are hundreds of videos on Ian Chow-Miller's YouTube page, many of which help illuminate a lot of the topics covered in this book.

LEGO Education Community

community.education.lego.com

LEGO Education has provided this community forum to help its users question, explore, share, and communicate their ideas with each other and with LEGO Education employees.

LEGO Engineering

www.legoengineering.com

Hosted by Tufts University Center for Engineering Educational Outreach (CEEO), this site contains hundreds of articles pertaining to the world of robotics education. The author of this book is a contributor.

LEGO MINDSTORMS Blog

www.thenxtstep.com

This blog didn't change its name when LEGO created the EV3 line, but it is very current with information related to the EV3 as well as other issues in robotics.

STEM Education With a Robotics Implementation

stemrobotics.cs.pdx.edu/node/291

This site provides curriculum for robotics classes. The curriculum is free and customizable and anyone with an (free) account can add to it or remix different parts into their own.

Product Websites

Edison
meetedison.com
A newer entry into the educational robotics market, the Edison robot has built-in sensors, but no external ones for attachment.

HiTechnic
www.hitechnic.com
One of the companies that makes third-party sensors for EV3 robots.

LEGO MINDSTORMS
www.legoeducation.us
In the USA, this is the site for purchasing all LEGO Mindstorms robots and related products.

MindSensors
www.mindsensors.com
One of the companies that makes third-party sensors for EV3 robots, including a line of cameras.

Vernier
www.vernier.com
One of the companies that makes third-party sensors for EV3 robots, specifically for scientific inquiry.

VEX Robotics
www.vexrobotics.com
This is the site to purchase all VEX Robotics materials including the EDR and IQ lines.

Index

Page numbers in **boldface** are illustrations. Entries in **boldface** are glossary terms.

About the Author

Ian Chow-Miller is a New York native who has lived in Tacoma, Washington, for the past eight years. He began his career as a social studies teacher but switched to robotics a decade ago and hasn't looked back. He has written curriculum for robotics and trained teachers around the country. He is a member of the LEGO Educator's Advisory Panel and is a constant contributor to Tufts University's LEGO Engineering website. Ian has coached FIRST LEGO League consistently since 2004, and when robotics season ends, he starts coaching soccer. He is married to a wonderful wife and has two awesome sons who are budding engineers.